COLLINS · TRAVELLER'S GUIDE

WILDLIFE OF THE ARCTIC

RICHARD SALE & PER MICHELSEN

T0187298

DEDICATION

To the continuing survival of the creatures of the North

William Collins
An imprint of HarperCollinsPublishers
1 London Bridge Street
London SE1 9GF

WilliamCollinsBooks.com

First published by William Collins in 2018

24 23 22 21 20 19 18
10 9 8 7 6 5 4 3 2 1

A catalogue record for this book is available from the British Library.

ISBN 978-0-00-820556-0

Collins uses papers that are natural, renewable and recyclable products made from
wood grown in sustainable forests. The manufacturing processes conform to the
environmental regulations of the country of origin.

Printed in China by RR Donnelley Asia Print Solutions

Contents

Picture Credits

All photographs are by the authors with the exception of those listed below.

Opposite *Cape Tegetthoff, Franz Josef Land.*

Below *Magdalenafjord, Spitsbergen, Svalbard.*

Preface

In 1829 John Ross sailed to the Arctic. It was not his first trip north. On that first trip, an early British Royal Navy expedition to find a North–West Passage, he had retreated when apparently confronted with a range of mountains blocking the way east: he had probably seen a mirage of the type that was to deceive later explorers, but others on the trip urged him to continue. Back home his failure to do so led to vilification for a lack of gumption and Ross was ignored as a commander on later expeditions. The 1829 trip was privately financed, which explains his command, but the new trip went no better than the first. Ross was forced to overwinter in the ice, something he had avoided on his first trip. On that trip Ross had met Greenlandic Inuit, famously wearing full dress uniform to greet the skin-clad locals. On meeting the Inuit again, Ross wrote that it was 'for philosophers to interest themselves in speculating on a horde so small, and so secluded, occupying so apparently hopeless a country, so barren, so wild, and so repulsive; and yet enjoying the most perfect vigour, the most well-fed health, and all else that here constitutes, not merely wealth, but the opulence of luxury; since they were as amply furnished with provisions, as with every other thing that could be necessary to their wants'. In that one sentence Ross encapsulated the lure of the Arctic for travellers from temperate regions to the south. The country was, as Ross contended, wild, but its wildness was also its beauty. Later travellers discovered a land that could not only be harsh and unforgiving, but a land of crystal, of silent cold, at times filled by the ghostly pale, trembling light of the aurora. Where the summer light was of breathtaking purity, but illuminated monochromatic scenery, white geese and swans on a black tundra, white ice on a dark sea, colour being rare and confined to summer. A land where the Sun, when it appeared after the Arctic night, could be cold and red and dishevelled, not the sun they knew. A land that seemed empty, with the people and animals being thinly spread so the loneliness could be awesome. Early travellers brought back tales of amazing creatures and of the endurance required of visitors, the Arctic becoming a land of inspiration and imagination. When Mary Shelley wrote her tale of Dr Frankenstein and the creature he created, she ends with the creature heading towards the North Pole.

The Arctic still inspires. Adventurers test themselves against it. Its wildlife still amazes when film and television show Earth's natural wonders it is always the polar regions that draw the biggest audiences. But today the Arctic is in retreat. Humanity's relentless exploitation of the Earth's resources in the pursuit of progress has altered the climate and threatens the ice and ice-living organisms. It is a cliché that the loss of a species diminishes us, but it is true nonetheless. Even to people who have never seen a Polar Bear its loss will be immeasurable as the bear is iconic, both defining and reflecting the Arctic.

This book celebrates the Arctic, exploring the natural history that has so inspired generations.

The Arctic

Defining the Arctic

At first glance it might seem ridiculous that the Arctic needs defining: the North Pole is well-defined, as is the Arctic Circle, so surely it is the area surrounding the Pole which is enclosed within the Circle? But in practice, while the Pole is certainly within the Arctic, the Circle is rather a poor definition of 'the Arctic'.

The Circle is defined by the fact that Earth's rotation around its own north-south axis is at an angle to the plane that defines its rotation around the Sun (the plane of the ecliptic), but is at an angle of 66.56°(rather than the expect 90°). The Chaldeans knew this because they had observed that while some stars were always visible, some rose and set at different times of the year. The Greeks systematised that knowledge. They knew the Earth was a sphere. They knew, too, that the Sun was overhead at noon at all altitudes between about 24° north and 24° south and would have known that this meant the rotation axis was at an angle of about 24°. The Greeks also realised that if they assumed the stars were set on the inner surface of another sphere, the celestial sphere, and the Earth was projected onto it, then the 24° circles passed through the constellations of Cancer, to the north, and Capricorn, to the south. On the Earth these circles are the Tropics of Cancer and Capricorn. The Greeks also realised that at 66°N the sun would be visible all day at midsummer and not visible at midwinter, and that at the Pole there would be six months of summer and six months of winter. The 66°N circle projected onto the celestial sphere passed through the constellation of the Great Bear – Arktikos: they therefore called the circle the Arctic Circle. Today the constellation is more commonly known by its Latin name, Ursa Major.

But before considering the Circle's usefulness in defining the Arctic, it is worth noting that the Greek idea of six months of summer and six of winter, though neat, is not bourne out in practice because of atmospheric refraction (the deflection of light by the atmosphere), which causes the Sun's image to appear above its true position by about 2½times its diameter. The Midnight Sun – the romantic name given to the phenomenon of the continuously visible Sun – is therefore visible at sea level for about 150km south of the Arctic Circle. The phenomenon also means that the Arctic summer lasts longer than the Arctic winter, by an amount that increases with distance north: at the North Pole the summer is about 16 days longer than the winter. Refraction can also cause the Sun to rise after it has set for the Arctic winter, or to rise early for the Arctic summer. One of the most extreme examples of the latter occurred during the 1596-97 overwintering of the Dutch expedition of Willem Barents on Novaya Zemlya when the Sun appeared almost two weeks before it was actually due to rise. Such images are usually distorted or broken.

As a definition, the Arctic Circle takes no account of the Earth's climate. To the west the influx of cold air and cold water chills North America, while to the east the North Atlantic Drift moves vast quantities of warm water to north-west Europe. This warm water, and the warm, damp air above it, has a huge influence on the climate of the region, particularly on the British Isles and Norway. North of the Arctic Circle in Norway there are large towns, and both industry and agriculture are possible. The effect of the Drift is less pronounced in Sweden and Finland, though both benefit to an extent, and it is eventually lost altogether in western Russia. The effect can be clearly seen if London is considered. It lies at the southern end of the UK, a huge city with an equable climate for human life. At James Bay in Canada, at about the same latitude, female Polar Bears give birth in dens excavated into the ice and snow.

After discarding the Arctic Circle in our search for a definition of the Arctic, it is instructive to ask how the Antarctic is defined. The Antarctic Convergence, where cold polar waters meet warmer waters from further north, provides a neat way of defining the southern polar region. But such an idea fails in the north: while Antarctica is a continent surrounded by an ocean, the Arctic is an ocean (albeit part frozen) surrounded by continents. The southern limit of pack ice fails as the edge is seasonal and it would, in any case, be difficult to extrapolate its position across the continents. No sea-based definition is possible. Considering land-based solutions, an obvious candidate is the tree line, the northern limit for tree growth. That certainly smooths out climatic differences, but it is not as precise and easily identified boundary as might be expected. Local geology and geography, as well as local climatic effects, play a role in defining

Midnight, 21 June, on the sea ice of Baffin Bay, with Baffin island to the left and Bylot Island to the right.

the habitability of an area. Ground elevation and aspect, drainage and soil composition all influence plant growth so that occasionally patches of forest exist, to the south of which there are areas, sometimes significant areas, of tree-free tundra. On paper the tree-line is a solid, immutable line, but on the ground it is rather more insubstantial, forming a band over which the transition from true boreal forest to true tundra occurs. In Siberia this transitional band can be as much as 300km wide.

And so, in their search for a definition, scientists turned to a temperature-based definition. The initial proposal was use of the 10°C summer isotherm, a line that links points on the Earth's surface at which the mean temperature of the warmest month of the year is 10°C. The isotherm has the advantage of being closely aligned to the tree-line. It is usually assumed that the factor limiting the northerly spread of trees is the cold: that is correct, but not in the sense that is usually inferred. It is not winter cold that is the limiting factor – in Siberia trees grow at a latitude that experiences the lowest winter temperatures recorded in the northern hemisphere. The limit is summer cold. In the Arctic summer there is abundant light, but the tree can only utilize this energy source if its cell temperatures are sufficiently high for the chemical reactions of photosynthesis to occur. Thus it is summer temperature – which must be high enough for a long enough period – that is critical to tree growth. As temperature is both easily measured and understood, the required extrapolation over water is limited, and the 10°C summer isotherm is closely aligned with the tree-line, a more-or-less tangible feature of the landscape, the use of this isotherm as a definition of the Arctic would seem ideal. But again there are drawbacks. The isotherm is poorly (often very poorly) defined across the intercontinental waters of the Arctic fringe, and it makes no allowance for winter cold. The former is not too much of a problem since the isotherm is a less valuable measure in the oceans, but the latter means that the place where the lowest-ever temperature in the northern hemisphere was recorded (at Oymyakon in Siberia) actually lies south of the 10° isotherm.

Despite these drawbacks, the isotherm has been adopted as a useful measure of the border between the Arctic and the subarctic by many specialists since it was first suggested in the late 19th century, though there have been attempts to address the problem of winter cold. The Danish scientist Morten Vahl suggested that 10°C should be replaced by the temperature V, where $V < 9.5° - (K/30)$, with V and K the mean temperatures of the warmest and coldest months of the year. For the Siberian forest, where the mean temperature of the coldest month might be -40°C the mean temperature of the warmest month would then be 10.8°C. The Swedish scientist Otto Nordenskjöld – nephew of the first man to sail the North-East Passage – considered that the Vahl formula did not adequately allow for winter cold and suggested a refinement, with $V < 9° - (K/10)$. Using this formulation for the Siberian forest the mean temperature of the warmest month becomes 13°C. The Nordenskjöld modified formula pushes the Arctic boundary south in Asian Russia and North America, but still excludes some areas that would be considered Arctic by the layman – Iceland, much of Alaska and northern Fennoscandia (Fennoscandia being the combination of Norway, Sweden, Finland and the

Kola Peninsula, and land immediately south of the White Sea in Russia). These exclusions seem anomalous for reasons other than common perception. For instance, although Iceland lies almost entirely south of the Arctic Circle it is north of the tree line (though whether the island's present tree-less state is a man-made rather than natural phenomenon is a matter of debate). This issue, and others, were addressed by the Arctic Council – a joint initiative of the Scandinavian countries (including Iceland), Russia and Canada – which, a few years ago, defined a boundary for CAFF, the programme for the Conservation of Arctic Flora and Fauna, which pushed the Arctic boundary well south, including not only Iceland, but extensive areas of Fennoscandia as well as the hinterlands of Russia and Canada, and much of south-western Alaska. However, the CAFF definition excludes Russia's Kamchatka Peninsula and Commander Islands, while including the Aleutian chain, a decision that is surprising.

In this book I have taken a pragmatic approach, the southern boundary of the Arctic being essentially defined by Nordenskjöld equation, but being pushed south to take in areas whose exclusion is inappropriate. In North America Churchill and Alaska's Denali National Park are included as they are likely destinations of travellers. For similar reasons Iceland is also included (though some Icelandic bird species are not as they are common in southern Europe and their inclusion would look very odd). In the Bering Sea the Pribilof Islands, the Aleutian island chain, the Commander Islands, the Kamchatka Peninsula, and the north-eastern coast of the Sea of Okhotsk are considered to lie within the Arctic as these too are now included in the plans of many northern travellers. And, of course, northern Fennoscandia is included – it would seem outlandish to tell those who have visited North Cape that they were not actually in the Arctic at all. The same would also be true for bird enthusiasts visiting Varangerfjord. By extension, the northern coast of Russia is also included. The Arctic limit, as defined here, is show in the map on p. 33.

Time at the poles

Because the day is 24 hours long and there are 360° of longitude, a traveller following the Equator finds local time changing by one hour for every 15° of longitude (though time zones are not always so rigorously applied). But if that same traveller is intent on reaching the North Pole this time change becomes increasingly meaningless. As our traveller moves north, lines of longitude crowd together, but the time difference between them remains the same, so the distance that needs to be travelled between them reduces. At the Equator the circumference of the Earth is about 40,000km, so the difference that needs to be travelled there for a time difference of 1 hour is around 1,670km. At the Arctic Circle this distance reduces to about 663km. At 85° it is down to 145km and by the time the traveller is within 1° of the pole it is a mere 29km. At the pole itself of course the distance has shrunk to zero and the heavily booted, probably down-clad traveller can circle the Earth as fast as he can turn circles and for as long as he can repel giddiness. At the pole the time is all times – it is the time the traveller is facing right now, but also the time faced with a turn of the head.

One result of this is the need for an agreed line at which the day changes – head west across it and today becomes tomorrow, head east and it becomes yesterday. That line, the International Date Line, was chosen to avoid occupied lands: it takes a more-or-less north–south course from the North to the South Pole through the Bering Strait and across the Bering Sea, deviating to ensure that all the Aleutian Islands, part of the USA, lie to its east, while Russia's Commander Islands lie to the west. For all travellers the line has comic potential, but because of the effect of decreasing distance for 15° of longitude the further north travelled, the Arctic traveller has the better deal.

Rock

The break-up of Pangaea (the supercontinent formed about 350 million years ago) began about 100 million years ago. The break-up was caused by plate tectonics, the most striking from a point of view of both the Arctic and the Arctic traveller, being the opening of the mid-Atlantic Ridge. Today in Iceland the traveller can stand within the separation boundary between North America and Eurasia, the island's volcanic activity evidence of the spreading nature of the Ridge. But other tectonic spreading ridges were at work, for instance in the Labrador Sea, the net effect being to separate not only the Americas from Eurasia and Africa, but to move

Rubini Rock, a magmatic sea cliff at Hooker Island, Franz Josef Land. The rock architecture has made the cliff a breeding places for hundreds of auks.

Greenland away from Eurasia, to the east, and Ellesmere Island, to the west. Initially land bridges still remained, linking Chukotka to Alaska across the Bering Sea, west Greenland to Ellesmere, and east Greenland to Fennoscandia, a bridge that included Svalbard. Within this ring of land the Arctic Ocean was separated from Earth's other oceans.

The northern extents of the land which surrounded the ocean were largely massive shields of Precambrian rock, some of the oldest rocks on Earth. In Canada's North-west Territories rocks have been dated, by examination of zircon crystals that formed as volcanic rocks cooled, to an age of a little over 4,000 million years, with other rocks near Nuuk in west Greenland, pillow lavas and fine-grained sedimentary rocks laid down in deep waters, being aged at 3,750 million years: in 2016 stromatolites (layered calcareous structures formed from mats of cyanobacteria) dated to 3,700 million years ago were found in south-west Greenland, the oldest life forms capable of harnessing the sun's energy by photosynthesis so far discovered.

Many of the rocks that overlay these ancient stratas are fossil-rich sedimentary forms, though throughout the area there are igneous rocks, evidence of the volcanic activity associated with plate tectonics. These formations are found on Svalbard (Newtontoppen, the archipelago's highest peak, at 1,717m, is a granite intrusion), Greenland, and on the sheer granite faces of the Auyuittuq National Park in southern Baffin Island. Volcanism is associated with the tectonic activity of an oceanic plate subducting (slipping under) a continental plate, the oceanic plates being denser and so lying lower on the Earth's mantle. Subduction creates a zone where both earthquakes and volcanoes occur. Earthquakes are produced by plate distortions which cause fault lines, or when the subducting plate drags the overlying plate until a quake occurs to restore stability. The 'Pacific Ring of Fire', created by the subducting Pacific plate has examples of both, with volcanoes and the sites of earthquakes lying within the Arctic as defined here. The volcanoes of Kamchatka are superb examples of the form, while the Alaskan earthquake of 1964 both raised and lowered land – a wrecked ship was raised 9m in one place.

Snow and Ice

But while geology creates the basic pattern of continents and islands that define the Arctic, it is snow and ice which have shaped the land and which, in public perception, define the area.

Water is a curious compound, disobeying the standard three-phase behavioural pattern of substances as they change from solid to liquid to gas (and vice versa). Water can change directly from its solid form, ice, to a vapour directly by a process called sublimation, and can change the other way as well, a process called hoar-frost. Hoar-frost can be seen outside the

Arctic, though often frosts observed in more temperate zones are frozen dew, and is best observed when spidery patterns appear on cold window panes. In the Arctic the process creates one of the most beautiful of spectacles, that of diamond dust, ice crystals, typically less than 0.2mm across, that glitter and shimmer as they fall to the ground on cloud-free, sunny days when the temperature is low (usually at or below -30°C). Diamond dust can also be seen in moonlight, a rare, unearthly sight which should be savoured if encountered. Diamond dust is often the basis of parhelia, the curious arcs and haloes that can surround the sun.

Snow

Within clouds, small water droplets are swept upwards by rising air, becoming supercooled (i.e. the temperature of the droplets falls below water's freezing point), ambient temperature falling as height is gained. The supercooled droplets are triggered to freeze into ice crystals by the presence of minute dust particles, which act as nucleation centres, the accumulation of many such crystals then coalescing to forms snowflake. The beauty of snowflakes, which are usually about 1–2mm across, was described after naked-eye observations in the early 17th century by European scientists who noted their hexagonal (six-fold) shape, though the complex nature of the symmetry of individual flakes was not unravelled until the invention of the microscope. In 1665 the English scientist Robert Hooke published drawings from microscope observations, and 150 years later William Scoresby made a series of superb drawings of snowflakes he observed during whaling voyages to north-east Greenland. Scoresby noted that Arctic snowflakes were much more symmetric than those he had observed in Britain.

The six-fold symmetry derives from the way in which oxygen and hydrogen atoms interact to create a water molecule, the basic hexagonal lattice (known as 1_h by scientists) being responsible for the rare 3-sided and 12-sided snowflakes as well, and for producing the occasionally seen cubic/columnar forms. But despite the 1_h being symmetrical many snowflakes are asymmetric by the time they reach the ground, collisions and changes in atmospheric conditions destroying the pure form: almost all photographs of snowflakes are symmetric, either because they have been artificially produced in laboratories or, if 'natural', because the most beautiful, most photogenic ones have been isolated. It is always said that no two snowflakes are identical, a hypothesis which can never be proven. But as there are two isotopes of hydrogen (the second is deuterium which has a nucleus comprising a proton and neutron orbited by a single electron, rather than the common form with a single proton orbited by a single electron) and three stable isotopes of oxygen, there are six possible stable forms of water molecule. As each snowflake comprises many millions of molecules, the various forms varying the crystal structure in an almost infinitesimal, but cumulative significant way, it is probable that no two are identical.

Snowfall accounts for about 5% of the total precipitation that falls on the Earth annually, the remainder falling as rain. That does not sound like a great deal, but is amounts to more than 30 million million tonnes per annum, the bulk (about 90%) being seasonal – or temporary in temperate areas – with the rest contributing to the Earth's ice sheets. There are different forms of snow, these depending on density. Several things affect the density of snow when it is first deposited, but in general it has a specific gravity of about 0.1 (i.e. 10cm of snow is the equivalent in terms of water volume of 1cm of rain). Once deposited, snow is compacted: as individual snowflakes break up, the pieces pack close together. Compaction can occur entirely due to the snow's own weight if the snowfall is heavy, but can be aided by the action of wind, and by surface meltwater percolating down into the spaces between the snow crystals. These processes increase the specific gravity, the increase altering the characteristics of the snow. When the specific gravity reaches about 0.5, the snow is well-compacted and granular, a form well-known enough to skiers and mountaineers to have been given a specific name: *firn* in German and *névé* in French, though in each case the name can apply to snow with a range of specific gravities. The wind may affect the surface of such snow, often creating a crust fashioned into wave-like ridges, comparable in appearance to the waves the wind induces on the surface of a lake, a ridged surface known as *sastrugi*. Sastrugi can produce an uncomfortable ride for a traveller on a snow scooter if it is too hard to break through, the ride becoming a tooth-rattling, bone-shaking experience. But wind can also create a more level surface crust called wind slab, which can be very dangerous on slopes (and so is carefully watched for by mountaineers) as it can slide on the relatively unconsolidated base below it, causing an avalanche.

Sastrugi on sea ice of the Barents Sea.

Ice

There are three types of ice the Arctic traveller will encounter. One is glacial ice: formed from snowfall, this is not confined to the Arctic, being seen in mountain areas across the Earth. A second form, freshwater ice, is common to people living anywhere on Earth where the seasonal ambient temperature falls low enough for ponds or lakes to freeze. The third form is sea ice, which is confined to the polar regions.

Glacial Ice

As compaction increases, either due to additional snowfall or through an increase in local temperature that causes further surface-melting, the specific gravity of the ice rises further. Another process is also at work deep within the snow. Known as sintering, it involves the transfer of water molecules by sublimation: a complex, lattice-like structure is formed, with the air that formed a major constituent of firn being squeezed upwards, out of the lattice. When the specific gravity reaches about 0.8 the remaining air within the lattice becomes trapped in bubbles as escape routes to the surface are closed off. The snow has turned to ice.

As compaction continues and specific gravity increases further, the air within the trapped bubbles being compressed. Glacial ice usually has a specific gravity of 0.9. The specific gravity can increase still further, but not by much: the value for pure ice is 0.917, but glacial ice always has some trapped air.

The trapped air in glacial ice creates the phenomenon of blue ice. Air and other impurities in the ice scatter light at all frequencies, and as 'normal' ice is impure this means that it appears white. But water molecules preferentially absorb light with wavelengths at the red end of the spectrum. Pure ice, free of impurities, allows light to travel a relatively long distance, increasing the absorption of red wavelengths while allowing transmission of wavelengths at the blue end of the spectrum. The ice therefore appears blue, and the purer (and more compacted) the ice the bluer it appears.

Though ice usually appears as a very hard but brittle substance, glacial ice is actually a plastic material that will flow downhill under the influence of gravity (though the actual method by which glaciers move is complex, as we shall see below). In general glaciers gain mass in their upper region as the average annual temperature in that area tends to be below freezing, so that the mass accumulated from snowfall exceeds that lost through surface melting. In the glacier's lower region, mass loss exceeds accumulation: this is known as the ablation zone. Mass accumulation occurs solely as a result of precipitation, but ablation can result from melting for one of several reasons, or by direct mass loss. Melting can be from solar or geothermal energy input, or from friction due to sliding (though that is by far the least effective in terms of energy input). Direct mass loss occurs in tidewater glaciers where the collapse of the glacier front (or calving) creates icebergs. If the mass balance of a glacier is positive, i.e. accumulation exceeds ablation, the glacier grows. If the mass balance is negative the glacier retreats.

Having accepted that although ice can occasionally seem as hard and unyielding as steel is actually influenced by gravity, glacial flow would seem to be a straightforward process, but this is far from the case. The natural assumption would be that ice, which after all is well-known for being slippery, slides over the rock at the glacier's base. For this to take place, the basal ice of the glacier must not be frozen to the substrate. This may be the case if geothermal heat raises the temperature of the substrate, or if the basal ice layer's melting point is lowered due to the pressure of the overlying ice: even if the substrate temperature is sub-zero, it may still be higher than this lowered melt temperature. In such 'warm-based' glaciers the basal layer of ice melts and the glacier slides, with the meltwater acting as a lubricant. However, such glaciers are rare in the Arctic, most of the region's glaciers being 'cold-based', with the substrate being colder than the basal ice layer's melting point: in other words, the glacier is frozen to the substrate and basal sliding cannot occur. The advance of glaciers like this is by a process called ice creep. This differs from the flow of a liquid, as it is caused by the elongation and displacement of individual ice crystals. If a bar of ice is suspended in a room at a temperature below freezing point it will slowly elongate: that is ice creep. Under certain circumstances the bedrock may itself deform, contributing to glacial flow. In many Arctic glaciers the situation is even more complex, as the glacier may be cold-based in its upper regions and warm-based at lower altitudes. Recent measurements of the speed of some Arctic glaciers has shown that they are speeding up due to climate change, with increased surface melting causing a downward trickle of water to the glacier's base where, if it remains unfrozen, it acts as a lubricant, turning a previously cold-based glacier into a warm-based one. Faster glacial movement means that a greater volume of Greenland's ice sheet is now being discharged into the sea.

Although glaciers flow downhill, the ice will also pass over large impediments to its movement, such as large bodies of rock that form part of the substrate. The flow of the ice in such situations is complex, involving an enhancement of ice creep, but also regelation. Regelation is the melting of ice due to pressure and its subsequent refreezing when the pressure is released. The clearest example of this process can be seen in a cold room, if a wire with a weight attached is looped over a suspended block of ice. In time the wire will cut through the block and fall away, but the block will remain intact, with little trace of the wire's passage. Where the wire touches the ice, the pressure from the weight lowers the melt temperature. The ice melts, the wire descends through the thin water layer, and the water refreezes above it. The same process occurs in a glacier; pressure from the ice mass upstream causes the ice at an obstacle to melt. The water flows around the obstacle and refreezes on the downstream side. Ice creep is a slow process, so if the glacier's bed changes slope it cannot react fast enough to stop higher ice layers 'overtaking' lower layers. Fractures then develop in the ice as ice masses on differing slope angles split from each other. Where these faults reach the surface they form a series of parallel crevasses. Crevasses are usually transverse, i.e. they form at right angles to the ice movement, but other forms may occur: chevron crevasses are angled across the glacier and result from the effect of drag on the ice by the walls of a containing valley; while splaying crevasses, are a combination of the other two.

There are several types of glacier, but most are confined to mountainous areas. In the Arctic, though, there are essentially two types, valley (often called alpine) and piedmont glaciers. Valley glaciers are those confined to the valleys that the glacier itself has carved. Such glaciers terminate in a convex ice snout, which either gives birth to a river or, if the glacier reaches the sea, calves icebergs. Those that reach the sea are often called tidewater glaciers. Tidewater glaciers often carve their base to below sea level, and the sea fills the subsequent trench if the glacier retreats. This process is the origin of fjords: the most famous fjords indent Norway's coast. These are now essentially free of the glaciers that carved them, but elsewhere in the Arctic some fjords still retain their glaciers.

Tidewater glacier flowing down from Beerenberg volcano, Jan Mayen.

Piedmont glaciers are those in which the ice has reached a plain at the base of the mountains from which they form: the name means 'mountain foot' (and is shared by an Italian region, named after its position below an alpine mountain chain). Having escaped from the confines of its valley walls, the ice spreads on the plain to form a characteristic lobe.

Icebergs

Though icebergs are an oceanic feature, they are formed of freshwater ice, being calved from tidewater glaciers as the glacier front fractures into the sea. They therefore differ from Antarctic icebergs which, though also being freshwater ice, form when sections of ice sheets – glacial ice flowing above the waters of the Southern Ocean – fracture. The key difference between the two polar regions is temperature: Antarctica's ice is much colder than the majority of Arctic ice masses, and cold ice has a much higher tensile strength. The tides acting on Arctic tidewater glaciers cause the ice to flex and, ultimately, fracture, but Southern Ocean tides do not impart sufficient energy to overcome the strength of the ice.

Arctic ice shelves are smaller and rarer than their counterparts in Antarctica. They also form in a different way. While some are created by the flow of glacial ice across the ocean (e.g. the Milne ice shelf off Ellesmere Island's northern coast), others form when fast ice develops over many years. The most famous of the Arctic's ice shelves, that at Ward Hunt Island off the northern coast of Ellesmere Island (the starting point for many adventurers heading for the North Pole), was formed from such an accumulation of fast ice. Once an ice shelf has been created, snow accumulating on the upper surface and sea ice accumulating on the lower surface tend to thicken the shelf, while the accretion of sea ice on the ocean edge tends to elongate it. The seaward edge of the shelf may also break off, though the calved ice is rarely called a tabular berg, ice island, a specifically Arctic name, being more usual. Arctic ice islands comprise both freshwater and sea ice, whereas the tabular bergs of Antarctica are almost entirely freshwater ice.

As a consequence of the limited size of Arctic ice shelves few tabular icebergs are borne and these are small in comparison to the vast tabular bergs of Antarctica. Arctic icebergs, often misshapen in an aesthetically pleasing way, are smaller. But smaller does not mean trivial, as the *Titanic* discovered, with terrible consequences, on 15 April 1912. The *Titanic* did not, of course, have radar, though radar has its limitations as a tool for detecting icebergs: ice is only about 2% as reflective as metal and a misshapen berg scatters a radar beam in many directions. Today the International Iceberg Patrol (IIP) uses aircraft to spot bergs in the North Atlantic, the white berg being readily visible by eye, even on overcast days or in light fog (dense fog, of course, makes spotting impossible). There is no equivalent of the IIP in the North Pacific as icebergs are only calved in south-eastern Alaska: these are few in number and rarely escape the Gulf of Alaska.

The largest Arctic iceberg on record was 13km long and 6km wide. Icebergs more than 150m high have been observed, but they are comparatively rare. It is estimated that 30,000–40,000 icebergs calve annually from Arctic glaciers, though perhaps only 300–500 reach the open sea, as most are too small to survive for long periods. As ice is less dense than water, icebergs float with the larger fraction of their bulk (about 85% as a general rule) below the surface. This mass distribution represents a crucial difference between icebergs and sea ice. Although both ice forms are affected by ocean currents, icebergs are much less affected by the wind than sea ice (though if the above-water section of a berg is sail-like they will catch the wind). The difference can occasionally result in the bizarre vision of a berg moving in the opposite direction to the sea ice, and ploughing a course through it. Icebergs may also plough through the seabed, something which would seriously endanger oil pipelines and underwater cables.

Icebergs are eroded by a combination of sunlight, wind and wave action above the surface, and water temperature and wave action below. In time they disintegrate, often rolling over when differential erosion makes them unstable. Rolling bergs are extremely dangerous: people on them would be thrown into the sea and sucked under to almost certain death by induced currents, and boats beside them can be overwhelmed. As a consequence few scientists now ever land on large Arctic bergs and the rule for all travellers is stay well away. Overturned bergs can usually be spotted by the surface pattern created by underwater wave action. On rare occasions, icebergs will disintegrate explosively, but most die quietly, the fragments being given the banal name of bergy bits if they measure 2–5m across. Smaller chunks are brash ice, a name shared by broken pack ice. The expressive name 'growler' is given to a specific class of berg debris: growlers, which can also be produced by the collapse of a glacier front, are flat-topped masses that float low in the water. Large growlers can be a problem for ships as they can be easily missed by radar. Smaller growlers can be a problem for boats such as zodiacs, as if unseen they can be steered over, causing damage to propellers.

Freshwater Ice
As well as being able to transform directly from solid to 'gas' and vice versa, water is also curious in the way its density behaves as it cools. Initially the density increases, as would be expected. But it reaches a maximum at 4°C (the exact figure is 3.98°C for pure water) and then decreases with a further reduction of about 8.5% when it freezes. (As a digression, it is this decrease in ice volume that causes the damage to household pipes if water systems freeze during winter: as density increases, volume increases, the expanding ice rupturing a pipe or pushing a joint apart, but the damage is only apparent when the thaw occurs and water escapes.)

As ice is less dense than water, it floats. As the surface layer of a body of freshwater (a pond or lake) cools towards 4°C the water becomes denser and sinks, drawing warmer water to the surface. This process continues until the temperature of the entire water column reaches 4°C and maximum density is achieved. Further cooling of the surface water then causes it to becomes less dense, and this cooler, less dense water can form a stable layer on top of the denser water column. Now, if the air temperature above the lake is lower than 0°C ice can form on the surface. Freezing starts at a nucleation point and from it needle-like ribbons of ice may occasionally advance rapidly, an interesting feature of these ribbons being that they are eventually curtailed by their own success. The reason is latent heat, the heat absorbed or released by any substance as it changes phase (i.e. from solid to liquid, liquid to gas, or the reverse of each): when latent heat is absorbed or released, the temperature of the substance does not change – only the phase changes. As the ice ribbon needle advances across the water surface, the freezing water gives up its latent heat, warming the water ahead of the needle point, bringing further freezing to a halt.

As all who have stepped on the sheet of ice formed on a puddle after a frosty night will know, the first skin of ice on still water can be extremely thin, perhaps only 0.2mm thick: it is essentially two-dimensional. Because this thin surface ice skin acts as an insulator to the water below it, the process of increasing the thickness of lake ice is slow, further '2D' layers forming beneath the original. There is little downward (i.e. three-dimensional) growth, the junction between the water and the ice being very smooth – as anyone who has ever extracted a sheet of ice from a frozen lake will know, it is glass-like, a smooth-sided pane of ice. Usually the ice will also be as clear as glass: freshwater ice forms relatively slowly, and impurities are expelled from the crystals as it grows. If the ice forms very quickly these impurities remain trapped in the lattice and the ice is opaque, but this is rare. Fish survive in lakes with surface ice layers as they can still swim and feed as oxygen in the water is replenished by inflowing streams. However, in ponds the situation is different. If the pond is shallow, ice can eventually extend all the way to the bottom, entombing (and, obviously, killing) fish and other aquatic organisms. If the pond is of limited volume the surface ice prevents oxygen absorption and the dissolved oxygen may become exhausted, the pond's inhabitants dying by asphyxiation.

Rivers differ from lakes and ponds in two significant ways (with regard to ice formation). Water flowing downhill converts potential to kinetic energy, mitigating against freezing, while the tumbling action of the water created by gravity and a rough stream bed causes the cooler surface water to depth and so acts against the entire water column reaching a stable temperature. Large rivers rarely freeze entirely, though it is possible for ice to form at any point through the water column because of the tumbling action, this ice often plating out on cold surfaces at the river's edge. It is then possible for the ice to grow outward from the edge and even, if the river is slow-moving or narrow for this ice to meet so that the ice layer extends across the entire river width. As rivers tend to be slower towards their mouths, such ice may migrate upstream: it is this ice which breaks into chunks during the spring thaw and which can then create ice dams closer to the mouth.

Frozen lakes and rivers are a hazard for the Arctic traveller, and extreme caution must always be exercised. A good general rule is to always cross a river at its widest point. This

seems counter-intuitive, but it minimises the risk that the ice thickness has been eroded by sub-ice river flow. On lakes, do not walk one behind the other: if the first walker causes the ice to crack it may not break until the second man arrives. In general, 5cm of freshwater ice will support a man, but as with all general figures there are parameters that can affect the strength of the ice and the figure should not be taken as anything other than a guide.

Sea Ice

Although glacial ice has carved the land masses of the Arctic, it is the sea ice which, in both the popular imagination and in terms of area, defines the region. The central Arctic Ocean is covered in perennial sea ice, at the fringe of which is further seasonally varying ice cover.

Due to its salinity, sea ice forms in a very different way to freshwater ice. The Arctic Ocean has a lower salinity than other oceans due to the inflow of the huge continental rivers of Asia and North America (although the Arctic Ocean has only about 1% of the Earth's volume of seawater, it receives around 11% of the total freshwater input) and because the rate of evaporation from it is much lower than in other oceans, being zero where ice coverage is complete and much lower at the ice edge because the rate the rate of evaporation from cold water to a cold atmosphere is lower than in warmer oceans.

The presence of salt (chiefly sodium chloride, which constitutes 78% of the salt burden, with magnesium chloride (*c.* 11%) and other salts at lower concentrations) lowers the temperature at which water freezes to between -1.8°C and -2.0°C. It also changes the way in which the water freezes. Sea water does not exhibit a maximum density at 4°C, the density continuing to rise as the temperature falls. So as the surface layer cools it sinks, a situation never being reached where cool water floats on a layer of slightly warmer water. The whole water column has therefore to cool to its freezing point before ice can form. However, this does not mean that the entire Arctic Ocean must freeze solid to the bottom at once: a water column depth is dependent on salinity, and there are sharp discontinuities in salinity in the ocean that effectively demarcate these water columns. There is also a further effect of salts being lost from the surface layer of water as the temperature falls: this leaching leads to a surface layer of lower salinity (and therefore a higher freezing point) that aids the freezing process. As an aside, the leached salts increase the salinity of the water beneath the ice, helping, in part, to drive the Atlantic conveyor, a thermohaline current which is, in part, responsible for the Gulf Stream/North Atlantic Drift which pushes the Arctic to the north in the eastern Atlantic and whose cessation has been suggested as one possible outcome of global warming.

In the first stage of sea-ice formation, crystals (usually plates or needles 3-4mm across) grow to create frazil ice. The crystals then multiply to form a greyish surface layer that behaves a little like a thick, syrupy liquid that is known as grease ice. The crystals in this thick soup now coalesce, forming plates that initially remain flexible enough to bend and move with the action of waves and winds. If there is little wind and a calm sea, the plates may form extensive sheets known as nilas ice, but in more turbulent seas the plates thicken and collide, forming roughly circular plates, each with a raised edge caused by the rubbing against other plates: this sea ice form is called pancake ice. Eventually the plates coalesce to form a continuous sheet that thickens as it ages. The thickening process is also dependent on sea conditions: in turbulent seas ice usually thickens through the accumulation of frazil ice crystals on the lower surface, but in calmer seas long, columnar crystals may form.

Whether the thickening of the ice is by frazil or columnar crystals, it is, of course, temperature dependent. If the air temperature is very cold the ice may thicken by up to 20cm daily, though this rate inevitably declines because the thicker the ice becomes the more it acts

Above *Nilas sheets formed on a calm sea.*

Left *Pancake ice.*

as an insulator. As a consequence, thickening rates are rarely more than 40cm in a week or more than 2m in a year. Seasonal ice is therefore usually about this thick, though 'old' or 'multi-year' ice can be up to 8m thick (though 4–5m is more usual). This immense thickness derives not only from freezing of seawater on the lower ice surface, but the accumulation of snow on the upper surface, this compacting to form new (freshwater) ice. In general, sea ice of about 20cm thickness will take the weight of a human (compared with the *c.* 5cm quoted above for freshwater ice) – but again this should not be taken as a golden rule, as there are many factors that can potentially weaken the ice sheet. Since much of the salt is leached from the surface layer of sea ice, this thickness raises the question of why there is such a substantial difference in the strength of the two ice forms. The answer is entrained brine pockets which, due to their high salinity, do not freeze and which weaken the ice. Two simple experiments illustrate the existence of the brine channels and their effect on ice strength. A sheet of freshwater ice only 5–6mm thick can be handled as if it were a pane of glass, but to handle sea ice the same way requires a thickness of about 6cm (around ten times the thickness). More dramatically, if a 'pane' of freshwater ice is dropped it will shatter into a myriad of shards in much the same way as glass would. But a pane of sea ice will land with a splash, rather as a ball of treacle or a dropped ice-cream would.

Another rule of sea ice that is helpful for a traveller (but again not absolute) is that grey sea ice is thin and should be treated with caution (the colour being due to the dark sea visible through the ice), while white ice is thick and therefore likely to be stronger. However, snow falling on grey ice can turn it white …

Fast ice and pack ice

Sea ice can be anchored to the shore, forming what is called fast ice. Away from the shore sea ice is usually known as pack ice: to complete the picture, ice scientists define three forms of Arctic sea ice – pack, fast and polar cap ice, the latter being the permanent ice around the North Pole. Unbroken pack is the term for complete sea ice cover. However, heavy swells, the wind and currents can break up a continuous ice sheet, particularly during the Arctic summer when the ice thins due to melting. When the ice covers about 75–80% of the water surface it is often called close pack ice. Open pack ice refers to a coverage of 50–75%. There is no accepted term for coverage below 50%, ice floes being a term used to describe large sections of broken sea ice littering the water. As the floes are broken up by wave and wind erosion, or by collision with other chunks of ice, a mass of small ice pieces, brash ice, is created. Occasionally a vast expanse of sea ice will be eroded at its edge, so that rather than forming an area of closed or open pack, sections break free and are quickly smashed into brash. This abrupt edge, which can look as clean cut as the division between land and sea at a temperate shore line, is often called (a little confusingly) the floe edge.

The attachment of sea ice to the shore – the fast ice – is often so tenacious that movement of the pack causes a fracture between the pack mass and a section of fast ice. The fracture line often runs more or less parallel to the shore, the pack retreating to leave an area of open water called a lead. Leads may also form where the pack fractures due to currents or wind erosion. Leads are a nuisance to travellers on the sea ice: they often require long detours to a point where the lead narrows sufficiently to allow a crossing, the only alternative being to wait until the lead freezes over. Leads can be hazardous to boats, even to ships, and to Arctic cetaceans as they can form relatively quickly and close just as swiftly, trapping and crushing boats or marooning whales far from open water. Whales have often been seen desperately keeping a breathing hole open in a closing lead, in the hope that an escape route will open. Polar Bears may 'fish' for Beluga in such situations. Each time a whale surfaces to breathe, the bear inflicts damage with teeth and claws until the exhausted Beluga is incapable of further dives and can be hauled onto the ice.

The fracturing of pack ice can also result in the formation of pressure ridges, as currents and waves force one section of pack to ride over another as a lead closes. Pressure ridges can also form in open or close pack ice if floes are driven against each other, a general freezing then creating unbroken pack with ridges in situ. Such ridges can reach 15m in height and, as with leads, are a considerable challenge for sea-ice travellers. As well as a ridge on the pack surface there is also a downward-pointing ridge below the ice, where the over-ridden flow has been pushed down. In shallow water this ridge can be driven into the seabed. Such downward ridges (occasionally given the tongue-in-cheek name of 'bummocks' – the opposite of 'hummocks') can be a hazard to drilling rigs or underwater pipes and cables.

Much more dangerous is the *ivu*, an Inuit word that describes a potentially lethal event in which a jumble of floes are pushed – by what process is still not absolutely clear, though it must involve strong on-shore winds or currents – at speed on to land, rather like a frozen tsunami. Ivu can kill, but are thankfully very rare. Scientists were initially sceptical of the existence ivus, but they have now been verified, with several being observed and studied.

Polynyas

Polynyas, the name derives from a Russian word for a forest clearing, are areas of open water within the sea ice coverage which may occur throughout the year, some in summer sea ice recurring annually. North Water in Smith Sound at the northern end of Baffin Bay and the Great Siberian Polynya in the Laptev Sea are classic examples. Annual polynyas result from reliable currents such as those

Leads in Pond Inlet, Canada. In the foreground fast ice is attached to the northern coast of Baffin Island. A lead has spilt the fast ice from the main pack. A second lead is seen snaking away towards Bylot Island, separating the main pack ice into two huge sections. It is spring, and the leads are starting the break-up of winter sea ice.

between islands, or from reliable weather patterns, such as off-shore winds. However, some polynyas exist away from such sources and are thought to be caused by upwellings of warm water, though the precise reasons for this are not fully understood. Polynyas are important ecologically, with the annual ones being critical for some species. The open water allows oxygenation, light penetration and the surface water to warm. The salinity of open water differs from adjacent waters beneath the ice. Gradients in salinity and temperature set up in the water create currents that take oxygen to the depths and bring nutrients to the surface, allowing organisms to thrive. The importance of North Water is reflected by the fact that the seabird population nesting within flying distance of it is measured in tens of millions.

Glacial Landforms
As well as carving the landscape, creating valleys and fjords (where over-deepened valleys reach the sea which then fills the valley when the ice retreats), glaciers produce features which remain to puzzle the traveller when the ice has disappeared: erratics, boulders deposited in areas of different rock type; *roches moutonnées* (rock sheep), boulders with one side shallow-angled and smooth, the other high-angled and rough, the difference caused by the direction of the abrading ice; and the various forms of moraine, fine rock debris abraded from the valley side or glacier bottom and formed into differing shapes – drumlins, eskers, kames and sandars – all of which add interest for the traveller and provide a habitat for plants and wildlife. Perhaps the most interesting features are pingos, ice mounds covered with soil, which may be up to 75m high and 500m across, and patterned ground. The latter are raised polygons of soil, sometimes with stones sorted by size which baffled early travellers as they looked man-made: in fact they are a natural feature of the annual freeze–thaw cycle.

Permafrost
While snow and ice are the most obvious examples of the effect of polar cold, the most widespread outcome is not immediately visible to the traveller. Permafrost develops where intense cold penetrates the ground, a vast area on the northern hemisphere, including land far south of the 'Arctic' being underlain by a frozen soil and rock layer. Permafrost, which may or may not include ice volumes, is defined as rock and soil in which temperatures do not rise above 0°C during two consecutive years. In Europe the influence of the North Atlantic Drift largely prevented the creation of permafrost, so Europeans only became aware of its existence when Arctic explorers attempted to bury their dead and encountered the unyielding, frozen ground beneath the shallow, seasonally thawed layer. Even after the discovery little interest was taken in the phenomenon in North America until the gold rush of the late 19th century began the economic development of the area, and the need to erect permanent structures. By contrast, the Russians knew of the existence of permafrost by the early 16th century, and writings from the 18th century include references to the remains of mammoths being found in the permanently frozen ground of Siberia.

The depth of the permafrost depends largely on the geothermal heat flux into the ground below the frozen layer and the net energy balance at the surface. In parts of Siberia, the permafrost layer is almost 1,500m thick: in North America it reaches depths of about 1,000m on Baffin Island and 600m on Alaska's North Slope. Such depths are almost certainly relics of the extreme cold of the Earth's recent geological history rather than a result of the present Arctic climate. One interesting feature of the distribution of permafrost beneath northern North America is that it is much more northerly to the east of Hudson Bay than to the west. The prevailing westerly winds in northern Canada blow across Hudson Bay; as the bay remains ice-free during the early winter, these winds pick up moisture, which is deposited as snow in northern Quebec and Labrador. Consequently snowfall in those provinces is much higher than in provinces west of Hudson Bay. The layer of snow acts as an insulating blanket, preventing cold penetration into the ground.

Strangely, ice itself, the 'progenitor' of the permafrost, also occasionally acts as an insulator, and it is thought that in some cases ice shielded the ground from extreme cold during the ice ages. As a consequence, the permafrost beneath most High Arctic glaciers is much thinner than in those areas of Alaska and Siberia that were not ice-covered during the last glaciation. The same is true under the sea, where unfrozen water insulated the seabed from extreme cold: permafrost occurs below the seabed only in areas where the shallow seas above the continental shelf froze solid, i.e. beneath the seas of eastern Russia and the Beaufort Sea.

Before permafrost was understood, building on it often led to disaster when cooking and heating produced a thaw, and structural collapse as here at Third Avenue, Dawson City in Canada's Yukon Territory during the Klondike Gold Rush.

In general, seasonal variations in the temperature of the permafrost do not occur at depths below about 20m. In summer the surface layer, known as the active layer, thaws. The depth of the active layer depends on the local energy balance; it may be as little as a few millimetres or more than several metres deep. Because the still-frozen permafrost beneath the active layer inhibits drainage, sections of the active layer may become saturated, forming an adhesive porridge (which rapidly accumulates in the tread of walking boots). Such areas are often called taliks. For the Arctic traveller in winter, taliks can be the cause of serious inconvenience as they may form hidden pockets that, if they lie beneath a thin surface crust, can overtop the boot of anyone unlucky enough to break through.

Aurora and Parhelia

Most travellers to the Arctic want to see the aurora borealis (Northern Lights) though as journeys in search of wildlife involve summer trips when the nights are short and are more often twilight rather than the total darkness in which the faint aurora is most visible, they are disappointed. That would appear to make mention of the Lights futile, but as they are primarily a polar phenomenon, a short description seems worthwhile, particularly as winter journeys northward are becoming more popular. The Lights are also entwined in the belief systems of native northern dwellers, adding an interest which extends beyond their ethereal beauty. For North American Inuit the Lights lit the path of the recently dead to the heavens, emanating from torches held aloft by ravens. The Inuit (and others) claim to occasionally hear the lights, the faint crackle they describe being identified as the feet of the dead walking across the crisp snows of heaven (there is no scientific basis for the Lights making any noise, but several earth-bound reasons why a noise might occur have been suggested). Greenland Inuit saw the lights as the souls of babies who had died soon after childbirth, peering down at a world they never knew and offering comfort to their parents. The Sámi peoples spoke of a 'fire fox' that raced across the sky each night, his coat sparking each time he clipped a mountain. They also wondered if, far away and out of sight, huge whales were spouting, with the jets of exhaled breath scattering the light of the stars. For the Chukchi of north-eastern Siberia the lights were the spirits of those who had died a violent death, perhaps from murder, suicide or childbirth. In the sky these wounded spirits played a suitably violent game, kicking and hurling a walrus skull around the sky. In New Zealand, the southern aurora was believed to be the reflection in the sky of fires lit by ancestors of the Maori, who had paddled their canoes south and become trapped in a land of snow and ice.

The aurora is created by the interaction of the solar wind, a stream of charged atomic and sub-atomic particles (amounting to about a million tonnes/sec) with the Earth's magnetic field. Accelerating in a spiral down the magnetic field lines the particles collide with the atoms of Earth's atmosphere, the collisions exciting (energising) electrons of the atoms of nitrogen, oxygen etc. As the electrons return to their ground state they emit the energy gained as light – green and red light from oxygen atoms, pale blue and violet light from nitrogen. Green light is the most common, the other colours being rarer. The form of the aurora varies, though there is only a small number of basic forms. In general arcs or veils are seen, these occasionally appearing as waterfalls of shimmering light. The aurora may appear to touch the Earth, but in fact the lower edge of the light is rarely below 60km, the aurora extending up to 400km above the surface.

Parhelia is the general term given to the range of solar and lunar haloes, arcs and sun dogs that occasionally surround the Sun, the effects being created by light refraction in ice crystals suspended in the troposphere. Most tropospheric ice crystals are either hexagonal flat plates or hexagonal columns. Light traversing these crystals is refracted at angles varying from about

22° to about 50°. As smaller angles are the most probable, the most frequently seen parhelion is a 22° halo surrounding the Sun. The minimum angle is 21.7° for red light, so the inner edge of the halo often appears red, with the remaining spectral colours becoming fainter (or being absent) as the halo fades away. Some parhelia are prismatic, the rainbow colours adding to the beauty and wonder of the sighting.

Avoiding Freezing

Ice cores from Antarctica and Greenland indicate that the Earth's temperature began to fall about 2½million years ago, and that a sustained period of cold occurred from 250,000 thousand years ago. During this period there were a succession of glacial maxima (the Quaternary Ice Ages), the last ending about 18,000 years ago when ice sheets extended across North America and Eurasia. While the Earth's temperature then rose, then fell briefly, it began to rise about 13,000 years ago, the Arctic as we now see it appearing as the ice retreated and plants and animals (the latter including human settlers) sought opportunities to expand their ranges. But this resettling did not occur in a single phase for all species. Throughout the Quaternary there were areas of land that were ice-free where species could still make a living, and as these areas changed from one interglacial to the next, this resulted in speciation, i.e the emergence of subspecies: Dunlin are the classic example of this, see p. 126. Even where the ice coverage was more or less complete, there were also peaks high enough to protrude through it. Called nunataks, these were areas where seeds, perhaps brought on the feet of birds or blown by the wind, might arrive. In most cases it is likely conditions were too hostile for plant survival, but in some they might have provided an early foothold, ready for range expansion when the local ice disappeared. It is also the case that seeds are hardy: seeds found in lemming burrows that had been frozen for thousands of years grew into healthy plants when given the opportunity. And as the greening of the lava fields of Iceland and Jan Mayen have shown (see p. 304) life is tenacious.

But to survive in the Arctic the plants and, when they arrived, the animals, needed to survive the harsh climate. The effect of temperature on biological function is reasonably straightforward. Organisms depend on enzymes to allow the chemical reactions that drive life processes to occur, and these proteins have optimal temperatures, i.e. temperatures at which they work best. At high temperatures enzymes denature (i.e. undergo structural change) and stop working. At low temperatures the enzymes do not denature but they do cease to function. In the absence of systems to combat changes in temperature the organism therefore has a limited temperature range over which it can survive (ignoring spectacular survivals such as those of nematodes, water bears and the eggs of some insects that can survive, in an inactive state, after a bath in liquid helium).

In the Arctic it is low temperature that presents the threat to the survival and this is a particular problem for ectotherms, i.e. those organisms whose temperature depends largely on the ambient temperature – fish, insects and amphibians – since endotherms (birds and mammals) can in principle increase their body insulation by adding blubber, fur or feathers, or increase their metabolic rate, assuming sufficient food is available. If the temperature of individual cells in an organism falls so that cellular water freezes, the ice crystals created can grow and damage membranes and other cell structures. In order to survive at low temperatures ectotherms have therefore had to evolve strategies to combat these

Coloured aurora, especially ones such as this which illustrates the full spectrum, are rare.

damaging and fatal effects. There are two strategies, freeze-tolerance and freeze-avoidance, though the two are not mutually exclusive, some species may use both or even switch between them. Freeze-tolerance utilises the fact that ice crystals must have a nucleus around which they can form. Such particles inevitably exist, but hydrophilic proteins are pumped out of the cells into the extracellular fluid. There they act as ice-nucleators, so the freezing occurs outside the cell. The proteins also order the water molecules so that crystal formation is slow, limiting local damage. Solutes concentrated in the cell also lower the freezing point of the cell fluids: glycerol is the most frequently found, and sugars are added to aid the prevention of damage to membranes and to maintain cell function. Freeze-tolerance is found in many invertebrates, including marine species and some insects (but is rare in vertebrates, though it does occur in some amphibians) and is effective in some species to temperatures as low as -70°C, though temperatures of -25°C to -40°C are more common.

Freeze-avoidance utilises the fact that very pure fluids, ones from which nuclei for crystal formation have been excluded, can be supercooled, that is cooled well below the normal freezing temperature. In this way water can be supercooled to about -40°C without ice formation. If solutes are added to the fluid so that the freezing temperature is further reduced, the strategy becomes even more effective. In some Arctic insects, anti-freeze compounds can amount to 25% of body weight. To reduce the amounts of particles that can aid ice crystal formation, organisms have efficient cleansing methods so that, for instance, they can empty their gut of food as residual particles can act as nuclei. Freeze-avoidance is practiced by insects and many spiders, and by more vertebrates – particularly polar fish – than practice freeze-tolerance. Though more energy-costly, as it requires the manufacture of antifreeze compounds, freeze-avoidance has the advantage of being more rapid in terms of switching from an inactive to an active state, and reduces the water loss that can arise through using freeze-tolerance. However, in general the lowest tolerable temperature is higher (usually in the range -5°C to -20°C) and the strategy is riskier: if intracellular freezing occurs it is very rapid and results in the swift death of the organism.

Plant adaptations
Plants exhibit a combination of freeze-tolerance and freeze-avoidance as well as a variety of insulation adaptations in order to survive low Arctic temperatures. Arctic plants grow close to the ground, though this low form can be either a genetically inherited morphology or one fashioned by the local environment. A classic example of the former is Arctic Willow, a dwarf willow species that grows as such even if transplanted to a less hostile environment. The low form of Arctic plants offers protection from the wind as well as allowing a more hospitable microclimate to be created. Wind desiccates the plant, a particular problem in winter when the frozen ground prevents water uptake. Because of friction, wind speeds are lower closer to the ground than they are higher up. Lower wind speeds also mean that abrasion from dust particles and snow is reduced. The depth of the active layer of permafrost also affects plant height: if the active layer is shallow (and so uniformly cold) and root growth is restricted, plant growth is slowed. Arctic travellers will notice that plants prefer to grow in sheltered places, but they may also be surprised to come across plants in isolated and exposed positions. The tenacity of life, here as elsewhere, is remarkable. The air temperature is higher closer to the ground, and the matted or cushion form of most High Arctic plants allows air to be trapped. Held close to the (relatively) warm ground, the warmer air helps to create a local microclimate of higher temperature. Many species show this convergent evolution, the most notable examples being Moss Campion and Purple Saxifrage. Moss Campion can occasionally be easily confused with a true moss, so tightly packed are the stems and leaves. Studies in the field have shown that in some cushion plants the internal temperature of the cushion can be as much as 15°C above ambient.

In winter the low form of Arctic plants also allows a covering of snow, which insulates the plant against plunging ambient temperatures. This is of particular value for species that are evergreen. Evergreen plants have the advantage of being able to photosynthesize in winter if there are bright days, and of being able to gain time at the start of spring by not having to wait until new leaves have grown. The new year's leaf growth is also protected during the early stages by the overtopping older leaves. While dead leaves at the top of the plant add insulation, those at the base are trapped close to the plant and ultimately add nutrients to the soil. Arctic soil is poor, and plants need to take advantage of any available nutrient resource:

Arctic travellers will often see relatively luxuriant plant growth beneath the breeding sites of birds where the ground is fertilised by droppings, or near an animal carcass. Some species also overwinter with well-developed flower buds to save development time in the short Arctic spring and summer.

There are many other adaptations. Leaves and stems, and the branches of woodier forms, are darker for greater heat absorption. Many species have hairs on stems, leaves and even flowers. The hairs on willow catkins not only trap air for insulation purposes, but the still air also reduces water loss. The parabolic shape of many flowers mimics that of solar furnaces, directing sunlight towards the plants reproductive structures to aid speedy development. The 'sun trap' that this induces within the flower proves a warm, welcoming microhabitat that attracts pollinating insects. Many Arctic flowers are highly phototropic, tracking the sun throughout the 24 hours of the polar summer day. The flower does not follow the sun by rotating – a recipe for disaster as following the midnight sun for day after day would twist the flower stem. The movement is created by the stem growing continuously, but always at a slower rate on the side towards the sun so the flower head tilts. Though the polar day is long, Arctic plants are able to photosynthesize at low light levels in order to take full advantage of available light.

Adaptations of terrestrial invertebrates

Though most Arctic invertebrates are aquatic (and chiefly marine), a surprising number inhabit both the tundra and the taiga edge. In terms of abundance the most numerous are worms (particularly nematodes and oligochaetes) and rotifers, with planarian worms being important within the taiga. There is also a diverse collection of freshwater copepods, and many insect species. Spiders are also important within the taiga. Almost all insect orders are represented in the Arctic (though there are very few beetles) and most are important as food sources for birds or as plant pollinators. In general, insects found in the Arctic are smaller, darker and hairier than their southern cousins. All three characteristics are adaptations to the northern environment. The smaller size is almost certainly because the insects are constrained by food resources (a smaller size also enables the insect to warm faster – but, of course, it also cools quicker) Arctic individuals also tend to be stockier, reducing their surface area-to-volume ratio, resulting in a relative reduction in both heat and moisture loss. The darker colour seems at first a contradiction of 'Gloger's Rule', which suggests that individuals tend to be white in the Arctic to aid camouflage. But that rule was formulated for vertebrates, and does not allow for the fact that in ectotherms the increased heat absorption (both direct radiation from the sun and reflected heat from the ground) of dark colours is generally more important than camouflage: a warmer insect can fly faster. The hairiness has arisen for similar reasons to the hairs of Arctic plants – the hairs trap air, reducing convective heat losses and providing insulation. Experiments have shown that if the hair is shaved from a caterpillar, the insect will lose heat by convection much faster than its unshaven siblings. It is difficult to know whether to marvel most at the ingenuity of the experiment or to pity the poor caterpillar.

Antennae and wing sizes are, in general, smaller than in southern species, to reduce heat loss. This is particularly noticeable in stoneflies, crane flies and some moths. In some species wings are entirely absent. Those insects that do have wings fly close to the ground where the air is warmer and wind speed lower. However, the wings of Arctic butterflies are not significantly smaller and are utilised as solar 'catchers'. Indeed, the basking strategies of butterflies are often useful in helping to identify species. Butterflies use one of three strategies – though individuals may not always limit themselves to just one. Dorsal basking involves flattening the wings so that the back and the upper wings are exposed to the sun. The butterfly leans towards the sun. This strategy is favoured by members of the Pieridae, the white butterflies (chiefly *Colias* spp. in the Arctic). In lateral basking the wings are raised so that the upper surfaces touch: the body is then turned sideways to the sun so the flanks and lower wing surfaces are exposed to it. Most fritillaries use this strategy. The wings may also be held in a V-shape, sunlight now being trapped between them, the wings acting as both collector and reflector. Members of the Lycaenidae, the blue butterflies, often adopt this strategy. The search for extra energy from the sun is also seen in insect larvae. Basking is practiced by some: the Woolly Bear caterpillar spends 60% of its active time basking, with just 20% spent eating. Mosquito larvae will move around a pool to track the movements of the sun, while the larvae of blow flies hatched from eggs laid on a carcass will be found in greater abundance and at a

Arctic Poppies at Hold-with-Hope, north-east Greenland. The poppies illustrate the typical parabolic flower and slender stems resulting from asymmetric growth, of all phototropic plants.

more advanced state of development on the southern side of the corpse.

Though these sun-seeking strategies are clearly effective, Arctic insects nonetheless have to complete their life cycles in ambient temperatures that are lower than those enjoyed by their southern cousins. They are therefore more active at lower temperatures. High Arctic mosquitoes go through their egg and larval stages in temperatures that may exceed 1°C only rarely, and Arctic travellers will note that mosquitoes continue to fly and feed at temperatures close to freezing, though they become much more active as the temperature increases. Yet despite being able to remain active at lower temperatures, many Arctic insects take much longer to go through their life-cycle than do related southern species. Some insects that range across both the temperate and Arctic zones have different life spans in different places. The springtail *Hypogastrura tullbergi*, for instance, has a life span as short as eight weeks in southern areas of its range, but in the High Arctic they can live for as long as five years. Though this difference may be extreme, there are many High Arctic insects with life spans of three or four years, and some High Arctic midges may take seven years to pupate. The most extreme case of such an extended span is the Arctic Woolly Bear caterpillar (of the moth species *Gynaephora groenlandica*), which takes at least seven, and as many as 14, years to develop to the pupal stage. In sharp contrast, the adult moth completes the life-cycle within a few days. The male dies after mating, the female after egg-laying: neither adult feeds. The reproduction of Arctic insects is, in general, as for southern species, but there are notable exceptions. In some species males are rare and females reproduce parthenogenetically (for example, some stoneflies, midges, black flies and caddisflies) Asexual reproduction allows spectacular and rapid population increases and so is favoured as it reduces the time for courtship, but it reduces genetic diversity. In the extreme case of the High Arctic black fly *Simulium arcticum*, the insects do not even go through the full life-cycle of egg, larva, pupa and adult, eggs developing inside the pupa and being released when the pupa dies, without an adult stage occurring at all.

Adaptations of aquatic organisms
As freshwater freezes at 0°C and sea water at about -1.8°C, aquatic organisms need in principle employ only freeze-resistance techniques adequate for a range of temperatures close to those figures. However, this logic only holds if the organism is in liquid water; for both freshwater and marine organisms situations may occur whereby they experience much lower temperatures, low enough to require more sophisticated freeze-resistance methods in order to avoid certain death. Large lakes and most rivers do not freeze entirely with fish and aquatic invertebrates surviving by employing freeze-resistance techniques and avoiding contact with the surface ice. Some river species may migrate to large lakes to reduce the risk of freezing. Invertebrates move to the depths to escape the advancing ice and may burrow into the lake bottom, though some insect larvae actually freeze into gravel beds close to the pond edges, using freeze-avoidance or tolerance to overwinter without ill effect. It is not, however, only the cold that has to be endured. With ice coverage preventing the absorption of oxygen at the surface, and the halting of photosynthesis, decomposition may cause an oxygen deficiency. Some species can actually switch from aerobic to anaerobic respiration (metabolising without the need for oxygen) to survive, but for many – both plants and animals – oxygen starvation results in death.

Most marine ectotherms use freeze-tolerance or freeze-avoidance techniques, but also rely, to a lesser or greater extent, on the fact that the freezing point of seawater declines as depth increases, so that migration to greater depth is an effective precaution against freezing. For intertidal animals (e.g. shellfish that graze algae at the tide level) the situation is very different as they may be exposed to ambient temperatures that are much lower than the freezing point of seawater. Such animals are invariably freeze-tolerant, able to survive if as much as 90% of their body water freezes. However, even for these remarkable survivors there is a problem if they remain above the sea ice, as prolonged exposure to the low ambient temperatures of the Arctic winter would almost certainly result in death. These animals must also avoid being caught where the sea is actually freezing, as during this time the ice is still in motion: moving ice is highly abrasive, with fast ice destroying most local life-forms during its formation. Being entombed within the sea ice as it forms would be equally deadly. Intertidal animals therefore migrate downwards as winter approaches. In rock pools sea ice formation results in salt leaching into the underlying water, raising its salinity and hence depressing its freezing point. Animals within these pools can therefore safely remain below the ice, often buried in the bottom sediment.

Arctic amphibians and reptiles
There are no truly Arctic amphibians or reptiles, but several species are found north of the Arctic Circle in the Palearctic. Only one species has been recorded north of the Circle in the Nearctic. In Fennoscandia and Russia as far east as the Urals, the Common Frog *Rana temporaria* is found to the northern coast and remains active to +2°C. The frog's range then becomes more southerly. The Moor Frog *R. arvalis* is even more cold-resistant. It is found in northern Sweden and Finland, on the Kola Peninsula and east to the Urals. In eastern Russia the Siberian Wood Frog *R. amurensis* breeds to the Arctic Circle around the Lena River and around the northern shores of the Sea of Okhotsk. In the Nearctic, the Wood Frog *R. sylvatica* breeds in Alaska as far north as Bettles, and in north-west Canada, where it has been found in the Mackenzie delta. This frog is capable of surviving short periods (up to about two weeks) at temperatures to -4°C. The Common Toad *Bufo bufo* has been found north of the Arctic Circle in Scandinavia. The Great Crested Newt *Triturus cristatus* occurs north of the Circle in Sweden. However, the most astonishing of all these amphibians is the Siberian Newt *Salamandrella keyserlingii*, the most widespread of all 'Arctic' amphibians being found on the southern Taimyr Peninsula and in central Chukotka. Adults are able to survive freezing to -40°C for extended periods and are active (though obviously sluggish) to +1°C. Most remarkable of all, newts excavated from depths of 14m in the permafrost have revived without apparent trauma. The species is very long-lived (probably up to 100 years) which may be a response to periods of prolonged freezing. The eggs of the species are known to survive short periods of freezing within an ice matrix. All these amphibians hibernate during the winter, choosing burrows or piles of rotting vegetation where they stay either singly or in groups. The Siberian Newt may hibernate for up to 8 months.

Two species of terrestrial reptile, the Adder *Vipera berus* and Common Lizard *Lacerta vivipara*, have been regularly recorded north of the Arctic Circle in Fennoscandia and western Russia. The Common Lizard is the more northerly of the two species, having been recorded to the northern coast.

Birds and Mammals
The survival strategies against the cold used by Arctic endotherms, or warm-blooded animals, are very different from those exhibited by the region's ectotherms. The simplest of these is adopted by the vast majority of Arctic breeding birds – they fly south, escaping the cold and reaching places where food is abundant. Several seabirds stay within the Arctic, feeding at the ice edge or in polynyas, but few terrestrial species remain. The Snowy Owl, the Lagopus grouse species, Gyrfalcon, Common and Arctic Redpolls and the Raven are often said to be resident throughout the winter. In practice all will move south if local food supplies fail; individuals that remain on their breeding grounds throughout the winter tend to be at the Arctic fringe and so do not contend with the full rigours of the Arctic winter. Nevertheless, these species must have adaptations to allow them to survive periods of intense cold. All Arctic birds have an increased feather density, while the Lagopus species and the Snowy Owl have feathered feet. The Snowy Owl also has modified foot pads to reduce heat loss, a characteristic it shares with the Raven. The owl's insulation is superior to that of the Lagopus species, but the gamebirds can reduce their metabolic rate, and also dig holes in the snow to gain shelter and the benefits of an

insulating snow layer, something the owl does not do. The redpolls also excavate snow holes, the ability of these much smaller birds to increase their feather density being limited, as there is always a trade off between insulation and efficient function.

As well as having a greater density, the feathers of Arctic birds form a particularly smooth outer surface, shedding the wind so that ruffling, with a consequent breakdown of the insulatory layer and increased heat loss, are reduced. Arctic species also have down feathers below their contour feathers, the down having a modified structure, being 'fluffier' so as to trap air and enhance insulation properties. In very cold conditions birds seek shelter from the wind and stand motionless, occasionally with one foot tucked into the body to reduce heat loss, or sitting so that both feet are covered. Foraging for food requires energy and if food is scarce and local conditions hostile, doing nothing may be the most energy-efficient strategy (though if the temperature is low enough so that the bird has to shiver to maintain warmth then the strategy may fail, more energy being required to shiver than to forage). As in marine mammals, the arteries of Arctic birds are enveloped by veins in the legs and feet so that venal blood is warmed by arterial blood, allowing heat to 'bypass' the feet, reducing heat loss. The temperature of the feet of some Arctic species may be 30°C lower than the body temperature. Systems like this are known as counter-current heat exchangers, and are common in Arctic animals.

Some mammals, e.g. Reindeer/Caribou migrate in winter, but for most the principle cold avoidance adaptation is to increase insulation or to avoid winter's cold by sleeping through it. Insulation can be added as fur, the main choice of terrestrial mammals, or blubber, the technique preferred by marine mammals. Polar Bears, which spends significant parts of the year both on land or sea ice and in the sea, use a combination of the two, while beavers and otters, which also spend time in water use fur only. Beavers and otters have a dense underfur that traps air as an insulator, and a coarser outer fur. This outer fur becomes wet and must be shaken to remove water when the animal emerges. Wet fur is only about 2% as efficient an insulator as dry fur, but these animals spend only relatively short times in water and so are able to maintain a dry underfur. The Sea Otter (*Enhydra lutris*), which spends virtually its entire life in sea water, also uses fur as an insulator. They have the densest fur of any mammal, with the hairs of the underfur reaching c. 125,000/cm², giving a total of c. 800 million on the entire body. This density allows the animals to survive in cold waters, but requires a great deal of grooming to remove the salt crystals that form within the pelt and to keep the fur in good condition. Sea Otters spend about 20% of their time grooming. Mother otters also frequently groom their young, the cub lying across the mother's chest, clear of the water. Sea Otters also have to maintain a very high metabolic rate to stay warm, eating up to a third of their body weight each day. They also blow into their fur or create water bubbles in order to enhance the air layer at the base of the pelt. This creates a layer of still air close to the skin: still air is a marvellous insulator (about seven times better than the rubber of a wet suit) and the best furs can create insulating layers that are about 60% as good.

Many species also have much thicker winter pelages to add insulation. The winter pelage of the Arctic Fox is not only thicker but is also white. This appears to be counter-productive as white is highly reflective, while a black pelage would absorb more of the sun's radiation, allowing the animal to gain 'free' energy. But white camouflages the animal against winter's snow, a fact that has led to the assumption that all white coloration serves the same purpose. Animals that change colour tend to be prey species (even if, as in the case of the Arctic Fox, for example, the species is also a hunter) and so camouflage is beneficial. But this simple theory does not stand close scrutiny, particularly as some species are white at all times, while the Raven, a truly Arctic bird, is black at all times. The UV vision of many animals has not been studied, but in those cases where it has, particularly in birds, UV sensitivity has often been found. It is, therefore, dangerous to make too many assumptions about animal coloration based solely on what we see as humans. So, overall, several notes of caution must be sounded when considering colour.

In common with other species (such as Musk Oxen), Polar Bears have guard hairs, longer and thicker hairs that add little to the insulation properties of the fur. The hairs of the underfur are wavy, which allows them to interlock when overlapped, trapping air, with the guard hairs protecting the integrity of the system when the animal is underwater. When the bear emerges from the water, the springiness of the guard hairs allows the fur to 'bounce' back and so encourages air entrapment.

The reduced efficiency of wet fur and the need for grooming mean that fur is not a comfortable option for pinnipeds and is out of the question for cetaceans. Nevertheless, pinnipeds do employ a combination of fur and blubber, though for most species the fur is sparse. Only for pups and the fur seals is fur a significant aid to insulation. The pelt of an adult Northern Fur Seal has a hair density about 35–50% of the density of the Sea Otter. Seal pups are born with a coat known as lanugo, this coat being shed when the pup has accumulated a blubber layer by feeding on milk that is super-rich in fat. Lanugo is thick as the pups rely upon it for insulation, and usually white for camouflage as most Arctic seal pups are born on ice. Interestingly, the fur seals and seal pups do not curl as an aid to staying warm (by reducing exposed surface area) in the way that terrestrial mammals do (the Arctic Fox gains a further advantage by wrapping itself in its luxurious tail). Seals need to be relatively inflexible to allow efficient use of their hind flippers, while their body shape, optimised for hydrodynamic efficiency, also acts against curling. Their blubber insulation is also so efficient that overheating is the more usual problem for a hauled-out seal.

Sub-cutaneous blubber has advantages for marine mammals as weight is much less of an issue, and not an issue at all for cetaceans. Indeed, for marine mammals bulk is an aid to keeping warm. Water is cold relative to body temperature, and it is highly conductive in comparison to air. Heat loss is proportional to body surface area, but thermal inertia (i.e. heat capacity) is proportional to body volume. Animals with small surface area-to-volume ratios can therefore maintain body temperature more easily. Blubber comprises lipids (fatty substances), collagen fibres and other connective tissues, and water. Cetacean blubber is rich in collagen fibres: they prevent the non-rigid mass from hanging, bag-like, from the animal, which would influence streamlining. The lipid content of blubber varies both from species to species and within an individual animal depending on age, position on the body and the time of year. In general, marine mammals have more blubber than is necessary purely for insulation purposes, the excess being an energy reserve. Centimetre for centimetre blubber is a poorer insulator than fur, but it has the advantage of being essentially maintenance-free. Though not as effective as fur, blubber is so efficient in keeping pinnipeds and cetaceans warm in water that the animals need a method of keeping cool. This is particularly true for pinnipeds, which haul out of the water to cool off. This seems odd to humans as for us cooling is far faster in water, but pinnipeds have counter-current heat exchangers, with arteries surrounded by a network of veins which work to reduce heat loss in the water, but can be used to increase heat loss in air when venal blood flow is increased. These systems explain why pinnipeds often wave their flippers and walruses turn red when they are hauled out.

One significant adaptation of Arctic mammals, both seals and terrestrial species, concerns the newborn. They experience a large temperature differential at birth, with as much as an 80°C change from the uterus to the outside world. Despite insulation this could lead to thermal shock. Newborns shiver to generate heat, but also employ 'non-shivering thermogenesis', a process unique to young mammals. In this, heat is produced by the metabolism of brown adipose tissue which is found in body cavities and around major organs. The colour of this tissue type derives from the mass of capillaries within it. Heat production by brown fat metabolism is about ten times higher than that of normal shivering. As the young animal develops, its brown fat deposits diminish, though a small amount is maintained in the adults of some species, the Musk Ox, for example.

The Arctic in black and white:
Arctic Fox in winter pelage ... and Raven.

A final common strategy in terrestrial mammals that overwinter in the Arctic is torpor, the state in which an endotherm reduces its body temperature to a new, lower norm, with metabolic processes slowing as a result, usually to about 5% of the rate at normal body temperature. Just as the normal body temperature is critical in maintaining body functions (meaning the animal must increase metabolic rate and take positive steps to avoid hypothermia, or suffer potentially lethal effects), the new, lower temperature is also critical, so that if the body temperature falls below it then the same response is observed. It is, therefore, not true to say that mammals that pass the winter in a state of profound torpor are independent of their surroundings – if they do not respond to a significant fall in temperature they become hypothermic and die, just as any non-hibernating mammal would. True torpor is confined to small mammals (primarily rodents, insectivores and bats: Arctic marmots seem to be at the upper weight limit for true torpor, with body weights of about 8kg) and a limited number of birds none of which are Arctic species. Before entering winter torpor fat reserves are laid down, but these may not be sufficient to survive the long Arctic winter and Arctic rodents prepare food caches, waking at intervals to feed. Once torpid, heart and breathing rates decline and the carbon dioxide level in the blood increases, reducing the metabolic rate. The animal falls into torpor relatively slowly, but arouses from it much more quickly. During early arousal the body temperature is raised without shivering, but eventually shivering does occur, this causing a rapid increase in body temperature. Field Voles (*Microtus agrestis*), from the fringe of the European Arctic, huddle together during winter, a strategy that may also be used by less well-studied Arctic species. However, not all Arctic rodents use torpor as a strategy. Lemmings, for instance, excavate burrow systems beneath the snow which, when lined with grass, allow communal living (with huddling augmenting the insulating value of the snow cover) while they remain active, feeding on sub-snow vegetation. The system is not without its drawbacks. Owls can hear the rodents even when they are moving below a substantial thickness of snow and will dive through to catch them.

The winter state of Arctic Brown and Black Bears differs from true torpor, and might more correctly be termed a state of winter sleep or dormancy. In these bears (and other large mammals such as beavers and skunks) the fall in body temperature is limited to perhaps only 2–4°C. However, the heart rate falls significantly (by about 80%) and the metabolic rate falls to about half that of the waking state. This would not be sufficient to prevent hypothermia unless the animal accumulated very large fat reserves prior to dormancy. Prior to winter's onset the bears gorge on a high-calorie diet. During the winter they lose some 25% of their body weight. The animals avoid dehydration during dormancy by neither urinating nor defecating. Female Polar, Brown and Black bears also give birth during winter dormancy. The cubs are very small at birth (at 0.4% of the mother's weight, the smallest of any mammal: human babies are around 5% of maternal weight). Even when they emerge from the birthing den, cubs usually weigh much less than 5%. Although female Polar Bears sleep the winter away, the males do not, though they may excavate a den in the snow in order to shelter during periods of bad weather.

The Future of the Arctic

Humans arrived in Arctic from Asia, crossing the Beringian land bridge which linked Chukotka and Alaska at a time when sea levels were lower as a considerable volume of water lay as ice across the northern hemisphere. In Asia these earliest Arctic dwellers had been Reindeer hunters, and then herders, but in North America they spread west at higher latitudes, where Caribou were absent and became hunters of birds and mammals, the latter including marine mammals. The history of the Inuit peoples of North America and Greenland is complex and deserves a book of its own. Suffice here to say that they were nomadic, moving as they exhausted the prey in a given area. They lived in skin tents, though winter dwellings, partly subterranean, were occupied for longer periods. Very few early signs of these people have been discovered, the most significant being the ritual site on Yttagran Island off Chukotka's coast. Here, perhaps a thousand years ago, a processional way of Bowhead skulls and ribs led to hearths where, it is assumed, whale meat would be burned as offerings so that the people's gods would bring the whales again: the Bowheads pass the island twice annually as they migrate to and from the seas beyond the Bering Strait. The Inuit clothed themselves in the skins of the animals they ate, but with a relatively small population and very effective, but limited weaponry, they had limited impact on species numbers. By contrast in Scandinavia and to an even greater extent in Russia, as weaponry developed and the population increased both the need for, and the ability to catch, fur-bearing animals increased. In Europe the Beaver was hunted almost to extinction, while as Russia expanded into Siberia Arctic Fox and Sable (*Martes zibellina*) were captured in numbers that almost defy belief. Sable was particularly valuable as its luxurious, silky fur is unique in being smooth in whichever direction it is smoothed, and because animals are found in different colours ranging from light brown to black: some scholars consider the Golden Fleece of Greek mythology was a Sable pelt as some animals may actually be that colour. In the case of Beaver, only the diminishing returns of hunting saved the species from extinction. In Siberia as the returns of hunting diminished, the hunters simply moved east into an apparently limitless land. When they finally reached the sea, they sent expeditions to search out new areas in which to hunt, finding, and almost destroying the Sea Otter and Northern Fur Sea. Here, again, only diminishing returns (and in the case of Sea Otters, luck) saved each species from total loss. In Canada the Hudson's Bay Company had a similar impact on the Beaver population (though its personnel did, in part, aid the exploration and mapping of the country). At the same time as hunters were denuding the forests and tundra of fur-bearing animals, the maritime nations were doing the same with the cetaceans leaving several, once numerous, species in a perilous state.

While the hunting of fur-bearing animals is now much reduced due to central heating systems and changes in people's attitude, and whaling is now heavily controlled (to the undisguised disgust of several nations) man has continued to have an impact on both the Arctic and its wildlife, both directly, with mining, oil and gas extraction, and even the dumping of nuclear waste, and indirectly with pollutants, which preferentially find their way north, and far more importantly, by warming the atmosphere.

The burning of fossil fuels has released vast quantities of carbon dioxide (CO_2) into the atmosphere. CO_2 is one of several 'greenhouses gases' (GHGs), others being water vapour, methane and nitrous oxide, which are so-called because they act on the Earth in a similar (but not identical) way to the glass in a greenhouse. The Earth's atmosphere is essentially transparent to short-wave solar radiation, but opaque to the long-wave radiation re-emitted by the Earth's surface. This re-emitted energy is absorbed by the atmosphere resulting in a warming. Without its atmosphere the average temperature of the Earth would be about -20°C: it is actually about +15°C. Ice cores extracted from Antarctic and Greenland have noted that this temperature has varied over time, as has the concentration of CO_2. Over the last 400,000 years the concentration has varied from 180ppmv (parts per million by volume) to about 280ppmv immediately prior to the Industrial revolution. The Ice Age history of the period

The ancient ritual site on Yttagran Island where Bowhead skulls and ribs were erected to form what appears to have been a processional way.

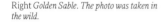
Above *Sea Otter. The species is a strong contender for title of the World's Cutest Animal, but that was no defence against fur hunters.*

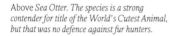
Right *Golden Sable. The photo was taken in the wild.*

follows in step: at times of low CO_2 concentration there is period of ice coverage, with high CO_2 levels there are periods of warming.

Some 30 years ago scientists began to report that the temperature of the Earth was rising. The temperature rise accelerated in the 1990s as a debate raged between those (chiefly scientists) who claimed it was a result of increases in GHGs in the atmosphere, these being the result of man-made emissions, chiefly from the burning of fossil fuels, and those who either denied the existence of the rise or claimed it was a natural occurrence, caused by changes in the Sun's output or other processes. Today the scientific community is essentially unanimous in believing that global warming is a result of anthropogenic effects, the evidence to support that position being clear. And nowhere clearer than in the Arctic where a series of feedback mechanisms means the area is warming at almost double the rate of the Earth as a whole.

The principal feedback mechanism is reduction of albedo as a result of sea ice loss. Albedo is the reflection of incident solar radiation by the white surface of the ice. Remove the ice and the dark sea below absorbs rather than reflects and warms the air above it. As the air warms, permafrost starts to thaw, trapped organic material then warms and microbial activity results in the release of both CO_2 and methane, the latter a 7,000 times more powerful GHG. As Earth warms vegetation spreads north: plant life is dark relative to the underlying soil and so albedo is reduced still further. As sea ice is lost, it becomes easier for ships and industry to move into the area. The Arctic is believed to hold very significant quantities of gas and oil, the burning of which will increase Earth's temperature further and so produce yet more sea ice loss. Although the loss is most apparent in the reduction of sea ice cover in summer, the warmer seas lapping at the edge of, and drifting below the ice are also thinning it: loss of ice volume aids both area shrinking and further thinning so that a vicious circle is set up. Some experts believe that the Arctic Ocean may be essentially ice free within 30 years. The increase in storminess seen over recent years is considered to be, in part, due to a loss of sea ice cover. And as the storms increase sea wave height and the frequency of large waves, and such waves batter the edge of the sea ice causing it to fracture and disperse, the storms caused by ice loss actually increase ice loss, another positive feedback mechanism.

Loss of sea ice does not, of itself, lead to a rise in sea levels, but the Greenland ice cap is melting. Were the sheet to melt completely sea levels would rise by 7m with catastrophic effects on coastal communities, and even entire low-lying countries – in Bangladesh much of the agricultural land is barely above sea level, on the Maldives, where the economy is based on tourism, the islands would sink beneath the waves.

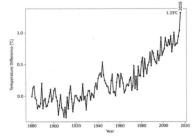

Increase in Earth land surface temperature since 1979. Drawn from data from NASA Earth Observatory. The sharp increase in 2016 was boosted by a strong El Niño event, though it is considered this contributed only about 20% of the temperature rise.

Spurred on by both the obvious signs of global warming and the overwhelming scientific evidence that is was anthropogenic in origin, the world's politicians have met many times to discuss the issues involved and to attempt a worldwide agreement to limit the Earth's temperature rise. The most recent was at Paris in 2015 by which time the CO_2 concentration of the atmosphere had reached 398ppmv, the 2014/15 winter temperature at the North Pole has been 5°C (35°C above the pre-1979 average) and the world had seen a succession of 'hottest year' on record (at the time of writing 2016 is expected to be yet another). An agreement was made, 195 countries stating their intention to limit Earth's temperature rise to 'well below 2°C above pre-industrial level' and 'pursu(ing) efforts to limit the rise to 1.5°C' with an accelerated reduction in fossil-fuel burning, increase in renewable energy sources and a carbon market for the trading of emissions. The agreement's aim is for zero emissions by 'the second half of the century' and a budget of $100 billion to aid poorer countries adapt. Agreement was made on 12 December, though it will not come into effect

Variation of ice coverage through the year, including summer 2017. The left scale is the sea area with at least 15% ice coverage. The black line is the average coverage for the years 1981–2010, with the grey shaded area representing ±2 standard deviations. The blue line is the coverage in 1979. As can be seen in that year the coverage was significantly higher in all months than the 1981–2010 average. The years from 2010 are represented by different coloured lines. The green dotted line is 2012: in that year the summer ice coverage was the minimum seen since records began in 1978. Data from National Snow and Ice Data Centre, University of Colorado, Boulder, USA.

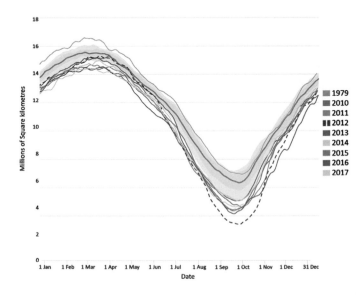

until signed by all parties during the period 22 April 2016–21 April 2017 (at the time of writing around 100 countries had signed). The agreement allows for stock taking every five years to ensure compliance. However, countries would be allowed to set their own targets (Nationally Determined Contributions (NDCs)) of emission cuts within a general framework. The NDCs should be 'ambitious' but sticking to them will not be legally binding. There was also no mechanism for forcing countries to comply though there will be a 'name and encourage' policy. Naysayers immediately noted that China, the US, India, Brazil, Canada, Russia, Indonesia and Australia which between them are responsible for more than 50% of GHGs, will be subject only to voluntary cuts without any penalty or fiscal pressure to comply. The agreement is also a fine example of circular logic in which the premise is contained within the conclusion – we will limit the temperature rise to 1.5°C because we will pursue efforts to limit the rise to 1.5°C.

In the aftermath of the Paris talks American president Barack Obama hailed the agreement, claiming it would create more jobs and economic growth, missing the rather obvious point that the pursuit of unfettered growth is what has brought us to where we are – an irony not lost on some experts dismayed by the lack of teeth in the agreement. The election of Donald Trump in late 2016 added to the dismay as he is a climate denier. His announcement of US withdrawal from the Paris agreement in June 2017 led to widespread condemnation, but many US States and corporations have declared their intention to continue abiding with the agreement's requirements. Crucially, the Chinese have also said they will continue to work in accordance with Paris. The major change would appear to be China's taking over world leadership from the US which is probably not what Trump had in mind.

To consider the plight of the Arctic and its wildlife in the face of the global problems associated with climate change might seem indulgent. Yet the fact that the temperature rise in the Arctic is more rapid than across the rest of the Earth means the region acts as a litmus paper, an indicator of what the future holds. The stress of environmental change will affect all Arctic species, but particularly the marine mammals for which sea ice is the habitat of choice. For the Polar Bear it may mean extinction or near-extinction in the wild, a tragedy that would touch the soul of millions.

Polar Bear tracks, Kvalvaagen, Svalbard.

Species Naming and Status

In 1963 Sir Peter Scott (son of Robert Falcon Scott – Scott of the Antarctic) and the driving force behind the founding to the UK's Wildfowl and Wetlands Trust) suggested the setting up of 'a register of threatened wildlife that includes definitions of degrees of threat'. That register is now maintained by the International Union for the Conservation of Nature and Natural Resources (IUCN) and is periodically published as a 'Red List', one of a series of threat categories, based on population and population trend being allocated to each species. Not surprisingly, given the sheer number of species on Earth, the first Red List category is Not Evaluated. All known mammals, birds, reptiles and amphibians have been evaluated, but the myriad of plants and insects, some of which have yet to be identified, are not yet included. Where insufficient data exist for an assessment of the risk to long-term survival of the species, the classification is simply Data Deficient. For those species where adequate data does exist, the category of Threatened species is sub-divided. The most extreme sub-category is Extinct where there is no reasonable doubt that the species has disappeared. Extinct in the Wild means the species exists only in captivity or has been naturalised away from its former range. 'Critically Endangered' means the species has an extremely high risk of extinction in the wild in the immediate future. 'Endangered' means a very high risk of extinction in the near future. 'Vulnerable' means a high risk of extinction in the near future. 'Near-threatened' species are at a lower risk of extinction and are of less immediate concern. A further category of 'Threatened' covers species with a lower risk again. All species are categorised as 'Least Concern' unless they are placed in a more onerous category. For further details of the otherwise ambiguous terms 'risk' and 'near future' the reader should consult IUCN Standards and Petitions Subcommittee 2016, *Guidelines for using the IUCN Red List Categories and Criteria Version 12*, available as a download at http://www.iucnredlist.org/documents/RedListGuidelines.pdf.

In the species accounts given below, the relevant category is given for all birds and mammals except where the category is Least Concern, this category being taken as read.

Trade in at-risk species is controlled by the Convention on International Trade in Endangered Species of Wild Fauna and Flora (CITES). The species covered by CITES are listed in three Appendices. Appendix I species are those threatened by extinction. Trade in these is permitted only in exceptional circumstances. Appendix II species are less vulnerable, but where unregulated trade is deemed incompatible with long-term species survival so that trade is strictly controlled. Appendix III species are protected in at least one country, that country having requested others to control trade. In the species accounts below the relevant CITES Appendix is noted. However, Appendix III species are not listed for non-Arctic countries.

The IUCN Red List includes the scientific name of all listed species. For birds the names reflect the latest expert view of the genus in which species should be placed. However, for some species this classification has not, as yet, been accepted worldwide. Anomalies include the re-assignment of Tattlers with the Tringa genus, and Broad-billed, Buff-breasted, Spoon-billed and Stilt sandpipers, together with Ruff and Surfbird to the Calidris genus. To avoid confusion, the more usual synonyms have been retained in this book. As the changes are likely to gain acceptance in the near future, travellers interested in the new classification are directed to the IUCN Red List, or to the *HBW and BirdLife International Illustrated Checklist of the Birds of the World*, del Hoyo, J., Collar, N.J., Christie, D.A., Elliott, A. and Fishpool, L.D.C. , Boesman, P. and Kirwan, G.M., 2016.

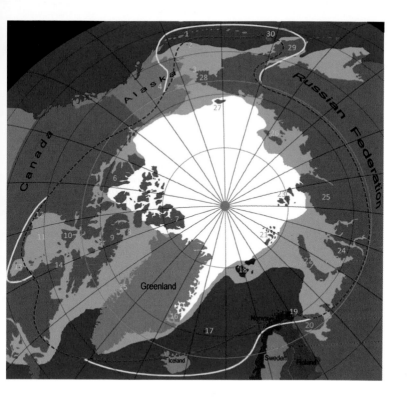

1	*Aleutian Islands*	16	*Scoresbysund*
2	*Pribilof Islands*	17	*Jan Mayen*
3	*St Lawrence Island*	18	*Svalbard*
4	*Denali National Park*	19	*Varangerfjord*
5	*Mackenzie Delta*	20	*Kola Peninsula*
6	*Victoria Island*	21	*Franz Josef Land*
7	*Ellesmere Island*	22	*Novaya Zemlya*
8	*Bylot Island*	23	*Severnaya Zemlya*
9	*Baffin Island*	24	*Yamal Peninsula*
10	*Southampton Island*	25	*Taimyr Peninsula*
11	*Hudson Bay*	26	*New Siberian Islands*
12	*Churchill*	27	*Wrangel Island*
13	*James Bay*	38	*Chukotka*
14	*Ungava Peninsula*	29	*Kamchatka*
15	*Disko Bay*	30	*Commander Islands*

The green line is the Arctic Circle. The red dotted line derives from the Nordenskjöld modified formula. The yellow lines indicate where the pragmatic definition of the Arctic boundary used in this book lies when it differs from the red dotted line.

Bibliography

The following books are recommended for further reference.

Byrkjedal, Ingvar and Thompson, Des, *Tundra Plovers*, T&AD Poyser, London, 1998.
Chandler, Richard, *Shorebirds of the Northern Hemisphere*, Christopher Helm, London, 2009.
Gaston, Anthony J. and Jones, Ian L., *The Auks*, OUP, Oxford, 1998.
Kear, Janet (Ed.), *Ducks, Geese and Swans* (Vols. 1 and 2), OUP, Oxford, 2005.
Macdonald, David and Barrett, Priscilla, *Mammals of Britain and Europe*, HarperCollins, London, 1993.
O'Brien, Michael, Crossley, Richard and Karlson, Kevin, *The Shorebird Guide*, Christopher Helm, London, 2006.
Olsen, Klaus Malling and Larsson, Hans, *Skuas and Jaegers*, Pica Press, Sussex, 1997.
Olsen, Klaus Malling and Larsson, Hans, *Gulls of Europe, Asia and North America*, Christopher Helm, London, 2003.
Pavlinov I. Ya., Kruskop S.V., Varshavsky A.A. and Borisenko A., *Terrestrial Mammals of Russia*, Izdatelvstvo, Moscow, 2002. (In Russian, not English translation).
Potapov, Eugene and Sale, Richard, *The Gyrfalcon*, T&AD Poyser, London, 2005.
Potapov, Eugene and Sale, Richard, *The Snowy Owl*, T&AD Poyser, London, 2012.
Potapov, Roald and Sale, Richard, *Grouse of the World*, New Holland, London, 2013.
Sale, Richard, *The Arctic: The Complete Story*, Frances Lincoln, London, 2008.
Sale, Richard and Potapov, Eugene, *The Scramble for the Arctic*, Frances Lincoln, London, 2010.
Waller, Geoffrey (Ed.), *Sealife: A Guide to the Marine Environment*, Pica Press, Sussex, 2000.
Wilson, Don E. and Ruff, Sue (Eds.), *The Smithsonian Book of North American Mammals*, Smithsonian Institue Press, Washington, 1999.

Parts of a Bird

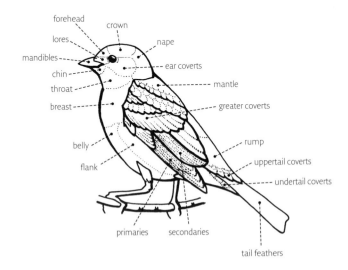

The Wildlife

Divers

With their delicately patterned plumage – seemingly the work of a talented painter rather than comprising individual feathers – the divers are among the most attractive of all northern birds.

For many, the haunting wailing or yodelling calls of the Great Northern Diver is also redolent of the wilderness, and it is no surprise to discover that these birds have become entwined in the myths of northern dwellers. In North America the call was thought to be the anguished cry of the dead calling for lost loves, an evocation that led to the belief that the birds guided the dead to the spirit world: sometimes the skull of a diver would accompany grave goods, carved ivory eyes replacing the originals, the better for it to follow the correct path. The bird's diving abilities also led to the belief that it could see in the dark and, therefore, could restore sight to blind people, the bird diving with the blind person on its back: it was believed the prominent white markings on the backs of the Great Northern (Photo 39a) and White-billed Divers (Common Loon and Yellow-billed Loon) were shell necklaces presented to the birds in thanks. In Siberia, native people incorporated the birds into creation myths, claiming that mud dredged from the bottom of the sea and brought to the surface on a diver's webbed foot began the process of building the land. Several Inuit dances included diver masks. Today, the Canadians have the Common Loon on their one-dollar coin, as a reminder of the wilderness that holds such a special place in the country's heart. Its appearance on the coin explains why, to the occasional puzzlement of first-time visitors from Europe, the coin is frequently called a 'loonie'.

Yet despite the reverence, the birds were killed for both meat and clothing by native Arctic dwellers. The dense feathers, evolved for keeping out the chill of Arctic waters for this essentially aquatic bird, allowed a diver carcass, snugly fitted to the head, to combat winter snowstorms: the clothing of 15th century mummies found in west Greenland included an inner parka, the hood of which was made from the skins of two Red-throated Divers. The waterproof skins were also useful for carrying the means to prepare fire, the bird being gutted to produce a bag: these 'loonie bags' are still occasionally seen in Nunavut. →

Red-throated Diver (Red-throated Loon) *Gavia stellata*

The smallest of the divers and, with a red throat that develops for the breeding season, one of the most attractive. The head and remainder of the neck are pale grey. The upperparts lack the chequer-boarding of the larger divers, being grey-brown with white speckling. In winter the grey neck and red throat are lost, the back and upper wings being covered in a myriad of white speckles: looking like the Milky Way, these spots explain the Latin name – *stellata* – stars (Photo 37e). The calls of the Red are also distinct from the voices of the other divers, being more waterfowl-like: in the UK's Shetland Islands the calls led to the bird being named the 'Rain Goose', though to be fair, given the rainfall of northern Britain most birds could be associated with its arrival. Being smaller, Red-throated Diver can nest on smaller lakes, though this often means that the local food supply is inadequate for chick-rearing, the birds having to make a large number of flights to gather food.

Circumpolar breeders, breeding on all the Arctic islands of Canada and Russia, though apparently absent from the New Siberian Islands. In winter the birds are seen in the North Atlantic, North Sea, Bering Sea and North Pacific.

a, b, d-f Red-throated Divers.
c Great Northern Diver.

Great Northern Diver (Common Loon) *Gavia immer*

The scientific name *immer* (Latin for immersion) derives from the apparently magical way in which the bird dives and disappears as it travels a long way underwater. A magnificent bird, with black head and a prominent red eye. The throat is black, ringed by a black-and-white striped collar, with broader areas of striping on the sides of the neck. The upperparts are checked black-and-white, the underparts white. The sexes are similar. In winter the red eye is lost, as is the chequerboard back and throat collar, the bird being a more uniform grey-brown.

Breeds across northern North America, including southern Baffin Island, on Greenland, Iceland, Jan Mayen and (irregularly) Bear Island. In winter they may be seen off both coasts of the USA, and off the coasts of the British Isles and Norway.

White-billed Diver (Yellow-billed Loon) *Gavia adamsii*

Status: IUCN: Near Threatened (probable rapid population decline, in part due to subsidence over-hunting).

The largest diver. The bill of the adults varies in colour from pale yellow to ivory (which allows both European and American names to be correct). It is also held slightly upturned, this accentuated by the upward angle of the gonys. Similar to the Great Northern Diver, but the chequerboard pattern of the upperparts has fewer, but larger, white spots. The white areas of the throat stripes are also larger. In winter the plumage is again very similar to the Great Northern Diver, though the bill colour is still distinctive.

Breeds along the northern coast of Russia to the east of the Urals, and on the southern island of Novaya Zemlya. In North America breeds in northern Alaska and the central Canadian Arctic, including southern Arctic islands. Seen off the north Norway coast, around northern Japan/Kuril islands, and off southern Alaskan and west Canadian coasts in winter.

→*Diver* is the British name for the five species of the genus *Gavia*, (from the Latin for 'seagull') birds so highly specialized for swimming that they have considerable difficulty walking. It is this inability that is believed to be the basis of the North American name – loon – for the birds, the word deriving from the Old Norse word *lømr*, lame or clumsy. The feet are webbed between the three front toes, with the legs positioned far back on the body, allowing the thrust from the feet to be developed behind the body for maximum efficiency when diving. The wings are not used for propulsion during dives, though occasionally deployed to aid fast turning: instead, the body is held rigidly, only the feet breaking the streamlined shape. The birds are superb divers, reaching depths of up to 75m, though much shallower dives are more normal. Dives usually last about 45s, though longer dives have been recorded. Underwater, the bird draws a transparent nictitating membrane across its eye as a form of 'contact lens' so as to retain excellent vision. The dagger-like bill is a highly efficient fishing tool. Often the birds swim with their heads submerged (Photo 41b) searching for prey before actually diving.

On land, as noted above, the position of the legs makes walking difficult and divers rarely travel far from water. On land the birds move by a series of inelegant hops or an equally inelegant shuffle, with the body held at an angle. The difficulty of locomotion on land means the birds take off from water. But they are heavy – White-billed Divers may weigh 6.5kg, and even the smallest diver, the Red-throated, weighs up to 2kg – and so take off after scurrying across the surface to obtain speed (Photo 37f). They then gain height slowly. As a consequence, divers look for long or large lakes. If the only available lake is smaller and tree-surrounded, it is not unusual for the birds to have to circle, gaining height, until they are high enough to clear the trees. Yet despite these problems, divers are strong fliers, often travelling considerable distances (up to 60km) to feed when they are rearing chicks, and making long migratory flights.→

a–c Great Northern Divers.
d–f White-billed Divers.

Black-throated Diver (Arctic Loon) *Gavia arctica*

Similar in appearance to the Great Northern Diver, but the head and nape are pale grey and the black, and striped patches of the next are larger. In addition to a yodelling call the birds also have a staccato, Raven-like, bark. In winter the birds lose the marvellous patterning. Nest on islets, at the lake shore and even occasionally in sheltered bays if the tidal reach is low.

Breeds in Fennoscandia, across Arctic Russia and in Kamchatka, with a few pairs also breeding in western Alaska. Siberian/west Alaskan birds are a subspecies (*Gavia arctica viridigularis*) with a green-sheen to the black throat. In winter birds are seen off the Japanese coast, and in the North Sea and North Atlantic.

Pacific Loon *Gavia pacifica*

Once considered to be a subspecies of the Black-throated Diver to which it is very similar. In breeding plumage the white throat stripes are less obvious, the head and nape paler. In winter the white flank patch of the Black-throated is diagnostic (Photo 41c). The two birds share habitats, habits and calls. Only in a few places in western coastal Alaska do the two species overlap during the breeding season: they may also overlap in winter off the coast of SE Alaska.

Breeds across North America and on the southern Canadian Arctic islands, and also in north-eastern Chukotka. In winter Pacific Divers are seen in the Bering Sea off Kamchatka and along the western US coast.

→The position of the legs also prevents the birds from landing feet first as other water birds (e.g. swans) do. In landing the diver resembles a seaplane, landing undercarriage-free on its underside. In flight the feet extend beyond the tail, a diagnostic characteristic (Photo 37d).

In winter divers are gregarious, but are solitary nesters and defend their territories aggressively against both rivals and potential predators of their eggs or young. Divers are monogamous, pairing apparently lifelong, though confirmatory evidence is scant for the larger species. Mating displays of the two larger divers involves little more than mutual bill-dipping, but the other three species include more elaborate courtship, the pair often making short, splashing runs with the body almost vertical and the bill thrust forward. Red-throated Divers can perform the most theatrical displays with side-by-side swimming, vertical calling, and long runs across the water with the bodies vertical, always with heads and bills thrusting forward, all in synchrony.

Divers eat a variety of fish species – freshwater during breeding (the Great Northern Diver has a penchant for trout in Iceland, though its Nearctic cousins prefer perch), marine at wintering quarters, though they will make do with just one kind if that is all their lake provides. They also eat amphibians and shellfish. The preferred nest site is an islet or floating mass of vegetation (Photo 37c and Photo 37b), but a marshy part of the lake shore will be used if nothing better is available. The nest, a pile of aquatic vegetation, is always close to the water because of the birds' poor walking abilities. There are usually two eggs, though one is not uncommon. The chicks are able to swim within 1 or 2 days (Photo 39b), though they often rest on their parents' backs (Photo 39c) or snuggle beneath the wings for warmth. Divers breed at two or three years of age. The birds are migratory, but do not travel far from the breeding territories.

a–c Black-throated Divers.
d–f Pacific Loons.

b

c

e

Grebes

Though superficially similar to divers, grebes have some distinctly different characteristics, suggesting a very different evolutionary path. The feet are not webbed, the toes being lobed to provide the paddles necessary for pursuit of prey underwater. The paddle stroke is also different from that of divers, a surface paddling grebe using the feet one at a time, and twisting them so that they move parallel to the surface. Lacking the divers' heavy bone structure grebes reduce their buoyancy by forcing air out from between their feathers to dive. As with the divers, the feet are placed far back on the body, making walking difficult (Photo 43c).

Apart from the flight feathers, grebes shed and replace their feathers continuously throughout the year, and have the curious habit of ingesting moulted feathers, and feeding them to their chicks. The birds also drink more than would be expected for their size, the water and feathers creating a paste-like mass that can amount to half the stomach volume. At intervals paste is regurgitated, leading to the suggestion that it allows the bird to rid itself of fish bones that might otherwise damage the lining of the digestive tract. However, other fish-eaters do not share the habit. The paste may also help the birds rid themselves of intestinal parasites.

Grebes have conspicuous courtship ceremonies. A pair of birds will stand upright on the water, breast-to-breast, with the heads turning from side to side (Photo 43b). They may then swim side by side, or even rush across the water side by side while remaining upright. In the 'weed ceremony' the birds dive together, each surfacing with its bill filled with weeds. They then stand facing each other, their heads moving sideways to display the weeds. The weeds may be used to build the nest, which comprises a heap of weed floating in the chosen pond or lake. The nest is anchored to aquatic vegetation (Photo 43d). Grebes hunt fish and aquatic invertebrates with short (usually less than 30s) dives at moderate depths of up to 20m.

Grebes are poor fliers in comparison to divers, the wings beating so fast the birds appear panic-stricken. As a consequence they are rarely seen in flight at their breeding territories. Nevertheless, the two Arctic breeding grebes are migratory, moving to southern coastal waters in winter. On migration they frequently fly at night. This has led to instances where in the early morning light exhausted birds have mistaken wet roads for streams and landed. They are then stranded, being unable to take off from land.

Slavonian Grebe (Horned Grebe) *Podiceps auritus*
Status: IUCN: Vulnerable (Population declining due to habitat loss).
Handsome birds when in breeding plumage, with highly conspicuous yellow 'horns' between a black crown and black throat. The neck and flanks are chestnut, the back pale grey. Sexes similar. In winter the horns and bright colours are lost, the birds becoming similar to Red-necked Grebes. Slavonians are smaller than Red-necked Grebes and are outcompeted by them: this probably explains why the smaller birds are found in a wider range of habitats. However, both birds may have been forced northwards by competition from larger southern species such as the Great Crested Grebe.

Circumpolar breeder, though absent from Greenland, and confined to more southerly areas of North America apart from the Mackenzie Delta. In winter found around the British Isles, in the North and Baltic seas, and off both American coasts.

Red-necked Grebe *Podiceps grisegena*
Less handsome than their horned cousins, but attractive birds with black crowns, white or pale grey faces and a red neck in breeding plumage. The upperparts are grey-brown, the underparts paler. Sexes similar. In winter the birds lose the bright colouration, being dull brown and white (Photo 43e). Highly territorial and very aggressive during the breeding season, Red-necked Grebes have been known to kill intruding ducks. If several chicks hatch the parents may split the brood when carrying them around.

Circumpolar breeder, though absent from Greenland and Iceland, and the High Arctic. Wintering birds are found off both coasts of America, and in the North and Baltic seas.

a-d Slavonian Grebes.
e, f Red-necked Grebes.

Tubenoses

Members of the order Procellariiformes, the albatrosses, shearwaters and petrels, share important characteristics. In particular they have the long, stiffened wings, tilted downward at the tips (Photos 45a, 45c). In soaring birds, such as vultures and other raptors, the wing-tip is upturned, this configuration giving greater stability but less manoeuvrability. The downturned wing-tip reverses this, allowing the bird to make quicker adjustments in the occasionally frenetic air streams close to the waves, but requiring greater control. All things are relative, however, and those who have watched Fulmars landing on nesting cliffs will know that the birds – much more impressive fliers when viewed from a ship – are nowhere near as comfortable as gulls when landing on the same nesting cliff.

Tubenoses have a lower body temperature relative to other birds, and a layer of subdermal fat (which adds insulation, but increases weight: the Procellariiformes are heavy birds). Most significantly, they have a highly distinctive bill form, with the nostrils housed in tubular channels set on top, structures which have led to the common name for the family birds – tubenoses (Photo 45b). The reason for the tubenose is still debated, though most authorities believe it is either an enhanced olfactory system, or acts as a pitot tube, measuring air pressure. In support of the former is the fact that the birds feed equally well during day or night: night feeding has advantages because although the birds can dive for food, they more normally feed by plucking food from the surface, and many prey species come close to the surface during the dark hours. In support of the idea that the structure measures wind speed is the birds' requirement to be able to 'read' the wind accurately because of their pelagic habitat. The characteristic flight pattern of the birds, gliding down with the wind and rising against it, is known as 'shearwatering', from which one group derive their common name. Shearwaters and petrels feed on fish, crustaceans and cephalopods.

Northern Fulmar *Fulmarus glacialis*

Superficially similar to gulls in its lighter morph, the Fulmar is distinguished at close quarters by the prominent tubenose and the fact that the suture lines of the bill plates are visible, and at a distance by the stiff-winged flight, the wings remaining stiff even during bouts of vigorous flapping. The bird comes in two basic morphs, dark and light (Photo 45d). Light morphs have pearly-grey heads and upperparts, with whiter underparts (Photo 45c). There is also a so-called 'double-light' (or L2), which is even paler. Dark morphs are darker grey above, though often with a paler head and tail, and pale grey underparts. There is also a darker form (the double-dark or D2) which is much darker overall above, though still with paler underparts (Photo 45f). Intermediate forms are also seen (Photo 45e). Male, female and winter plumages similar.

The distribution of the forms implies that the dark morph evolved in the North Pacific, the paler birds in the North Atlantic, when ice isolated an ancestral population. However, there are anomalies; there are many dark birds on Svalbard, and pale morphs in Chukotka. In general the number of dark morphs increases as the traveller heads north, which is also anomalous if the distribution hypothesis is strictly true.

Fulmar is circumpolar, breeding on sea cliffs. Pelagic in winter, moving ahead of the ice. In recent years there has been a marked increase in the population, probably as a result of the bird's habit of following trawlers to feed on fish offal. This has resulted in breeding sites becoming saturated and the species extending its range southwards. Northern expansion has been impossible as the range already extends to the ice edge. Fulmars feed as far as the ice edge, either by surface feeding or making shallow dives. A single chick is raised (two chicks are rare), this being fed by regurgitation, the adult's bill placed inside that of the young.

The name Fulmar is Icelandic, meaning 'foul gull' which almost certainly derives from the way the birds regurgitate fish oil when approached at a nest site. The oil is viscous and evil-smelling, but fortunately the bird's aim does not match its enthusiasm for regurgitation, and most oil jets miss their target.

a-f Northern Fulmars.

There are no Arctic albatross species, though the Laysan Albatross (*Phoebastria immutabilis*) which breeds on the Hawaiian islands may be seen as far as 55°N, though it is uncommon at all times and rare in winter. The Black-footed Albatross (*Phoebastria nigripes*) also breeds on Hawaiian islands and is relatively common off the southern shores of Alaska, though much rarer off western shores. The Short-tailed Albatross (*Phoebastria albatrus*), a very rare species with a population of perhaps no more than 1,500 breeds only on two islands off Japan's southern coast and has been seen on rare occasions off Alaska's south-western coast.

While no shearwaters or storm-petrels are true Arctic species, several breed in the subarctic. Shearwaters differ from Fulmars in having longer, narrower wings, narrower tails and longer, thinner bills. The species are named for their flight technique, the birds gaining energy from cross wave airflows as they 'shear' the wave tops, so close to them do they pass. It is thought that the origin of the name storm-petrel is the species' habit of sheltering in the lee of ships during storms, and the fluttering flight above the water of the feeding birds often involves them pattering across the surface with their feet, so giving the impression that they are walking on the water. 'Petrel' could then derive from St Peter, who also walked on water. An alternative name popular with sailors was Mother Carey's chickens. That name almost certainly derives from *Mater Cara* (Latin for 'Dear Mother' a synonym for the Virgin Mary) who was considered the protector of sailors. Although both names might suggest a reverence for the birds, this was not shared by everyone, as it is said that the birds were occasionally used as lamps, the oil content of their bodies being so high that a wick pushed down the throat and then lit would give a light that lasted for several hours. In addition to the species below, the European Storm-petrel (*Hydrobates pelagicus*) breeds on the Westmann Islands, off Iceland's southern coast, and a couple of places on the southern mainland.

Manx Shearwater *Puffinus puffinus*
Black upperparts, including back and wings, white underparts, including underwing. Male, female and winter plumages similar. Breeds on Iceland, but otherwise essentially non-Arctic. Wintering birds seen as far north as northern Norway.

Fork-tailed Storm-petrel *Hydrobates furcatus*
The only pale storm-petrel in the region. Pale grey, with a paler carpal bar on the upperwing and darker leading edge. There is a distinct black eye patch. The rump is very pale. The tail is deeply forked. Male, female and winter plumages similar. Breeds across the island arc from southern Kamchatka to southern Alaska, and winter in the northern Pacific and central Bering Sea.

Short-tailed Shearwater *Ardenna tenuirostris*
A sooty-brown bird with a paler, greyish panel at the centre of the underwing. Sexes similar. Breeds on Tasmania, but winters in the north Pacific and Bering Sea where huge flocks form. Flocks of more than 100,000 birds (perhaps as many as 500,000 on occasions) have been reported as far north as St Lawrence Island.

Leach's Storm-petrel *Hydrobates leucorhous*
Dark brown head with much paler face. The upperparts are also dark brown, but the upperwing-coverts are paler, creating a noticeable pale patch. White, V-shaped rump patch, though a dark-morph without this patch is found in the south Pacific. The underparts are uniformly dark brown. Tail deeply forked. Male, female and winter plumages similar. Breeds on southern Iceland, Newfoundland and across the island arc from southern Kamchatka to southern Alaska. Winters in the North Atlantic to central Norway, northern Pacific and central Bering Sea.

a, b *Northern Fulmars.*
c *Manx Shearwater.*
d *Fork-tailed Storm-petrel.*
e *Short-tailed Shearwater.*
f *Leach's Storm-petrel.*

Gannets and Cormorants

Historically these two families have been placed with the pelicans, but more recently, together with the darters and frigatebirds, have been considered a separate order, the Suliformes. Gannets and boobies are plunge-diving birds occurring primarily close to the Equator, the most northerly species being the Northern Gannet which breeds to the Arctic fringe in the Atlantic. There are no northern Pacific sulids.

Cormorants are a successful group of diving, fish-eating birds, well adapted to the aquatic environment, but much better on land than the divers and grebes, being able to stand upright and some even being able to perch and nest in trees. However, despite adaptations for a diving lifestyle, cormorant contour feathers are not completely waterproof, only the inner down layer preventing the skin from wetting. The wetting of the outer feathers may help reduce buoyancy, though that does seem a rather drastic solution. Even if correct, there is a price to be paid for the lack of waterproofing; the feathers must be dried after dives, resulting in the birds' characteristic pose with wings 'hung out' to dry (Photo 51d). This necessity means that cormorants usually fish close to land. It also means that most species favour tropical waters. Given the potential for cooling of the body when drying the wings, the fact that the misnamed Pelagic Cormorant is a true Arctic dweller is remarkable.

Cormorants have a long, flexible neck which aids the capture of prey. The bill is long and hooked at the end, while the tongue is rough, both adaptations to deal with slippery prey; the neck is pouch-like and can distend to allow fish to be swallowed whole, as they are after the bird has juggled them to ensure a head-first decent.

Cormorants are gregarious, both at fishing and nest sites. During the breeding season adults often acquire a white thigh patch and prominent crests and head plumes, together with coloured throat pouches, these features all being used in mating displays.

Northern Gannet *Morus bassanus*
Magnificent birds with white plumage apart from black wing-tips, a darker trailing edge on the wings, and a pale yellow head. Sexes similar. In winter the head is paler. Long, dagger-like bill is used to grab fish after a plunge, which can be from heights of up to 40m, water entry being occasionally above 100km/h. The nostrils open inside the bill to prevent water ingress when hitting the water. The eyes are well forward, giving excellent stereoscopic vision. Nests colonially, laying a single egg. The birds have no brood patch, so the single egg is incubated by the adult standing on it, using the blood vessels of the foot webs for warmth.

Northern Gannets breed on Iceland, northern Fennoscandia and Newfoundland/southern Labrador. Partially migratory to southern waters.

Pelagic Cormorant *Phalacrocorax pelagicus*
The smallest of the North Pacific cormorants. Plumage is black with distinct metallic green sheen. At close quarters the neck sheen may be seen to have a bluish tinge. Breeding birds have one or two short crests on top of the head, a small red *gular* (throat) pouch, prominent white thigh patches and may have white flecking on the neck. Sexes similar. In winter the birds lose the crests and thigh patches and much of the feather sheen. Despite the name, Pelagic Cormorants feed close to the shore (closer than the other Bering Sea cormorants), but in deeper water suggesting that they dive much deeper than related species.

Breeds on both sides of the Bering Sea, but much further north on the western side, reaching Chukotka and Wrangel Island; birds on Wrangel are the most northerly of all breeding cormorants. Breeds along the Aleutian chain. Southerly breeding birds are resident, northern birds migrate as far south as Baja California and Taiwan.

a-c Northern Gannets.
d Pelagic and Red-faced Cormorants on an islet off Unalaska, Aleutian Islands.
e Pelagic Cormorants in breeding plumage (taken in a southerly part of the species' range).
f Pelagic Cormorant in non-breeding plumage.

Red-faced Cormorant *Phalacrocorax urile*

The plumage is black with a purple-green sheen. The red face of the name is prominent and much more extensive than the breeding red of the Pelagic Cormorant. Breeding birds have one or two short crests on the head and a white thigh patch. Sexes are similar In winter the birds lose the crests and thigh patches.

Red-faced Cormorants breed across the island arc of the southern Bering Sea (including the Pribilofs) and on eastern Kamchatka and southern Alaska. Alaskan birds tend to be resident or to move only a short distance, but Russian birds may move as far as Taiwan.

Double-crested Cormorant *Phalacrocorax auritus*

The plumage is black with a subdued green sheen. The throat pouch is orange. Breeding plumes are the twin ear tufts of the name. These are black or grey in birds in the southern part of the species' range, but paler, even white, in northern birds. The Double-crested is the largest of the North Pacific cormorants. Sexes are similar. In winter the birds lose the crests. Work on the species breeding on Canada's Great Lakes was one part of the study of the effect of polymer microbeads – in this case beads ingested by animals at the plankton level finding their way through fish and hence into the birds – which has led to decisions to end their usage in cosmetics.

Breeds on both coasts of North America, as far north as Newfoundland to the east and to southern Alaska and the Aleutians to the west. Alaskan birds are resident, or may move short distances along the coast. Eastern birds move south.

Great Cormorant *Phalacrocorax carbo*

The plumage is black with a blue sheen. There are prominent white cheeks and throat. As with other cormorants, breeding adults have white thigh patches. They also have short black crests. Sexes are similar. In winter the birds lose the thigh patches.

Breeds on south-west and south-east Greenland, on Iceland, northern Norway and the Kola Peninsula. In North America breeds on Newfoundland and the adjacent mainland, but not on the Pacific coast. Some birds are resident, but others move south along the coast. All Greenlandic birds move to the south-west coast, but Icelandic birds are resident.

European Shag *Phalacrocorax aristotelis*

The plumage is black with a green sheen. The breeding crest is black and curved forwards. There is no thigh spot or prominent throat pouch in breeding adults. Sexes are similar. In winter the birds lose the crests.

European Shags are resident in western Iceland and on the Norwegian coast north to the Kola Peninsula.

a Red-faced Cormorant photographed at the northern extent of the species range, on the Pribilof Islands.
b Double-crested Cormorant.
c A photograph taken on an island off Kamchatka's eastern shore. The image is somewhat shaky as the author was perched precariously on an insubstantial ledge above a long drop into an angry Bering Sea. The upper bird is a Pelagic Cormorant in breeding plumage. The lower bird is also an adult – note the white patch – but curiously coloured. The extensive head coloration suggests a Red-faced Cormorant, but is yellow. The meeting seemed cordial rather than confrontational, though there is no evidence of hybridisation between the two species.
d Great Cormorant drying its wings.
e, f European Shags.
g Great Cormorants.

Swans

Wildfowl – swans, geese and ducks – are a successful group with more than 150 species spread across every continent except Antarctica. Their absence from the southern continent can be explained by the species' being chiefly herbivorous although there are also piscivorous ducks.

Wildfowl share some general characteristics. They are broad-bodied, long- (or very long-) necked aquatic birds. They have flattened bills with a horny 'nail' at the end, the edges of the mandibles having comb-like lamellae for straining food from water, and rasp-like tongues for manipulating food. In a few species, the sawbills, the lamellae are replaced by 'teeth'. Vegetarian wildfowl do not have bacteria in their gut to break down cellulose, so they gain nutrients only from cellular juices. The plant structure is broken down in the bird's gizzard, grit being ingested to aid the process. One consequence of this is that the birds have to eat a great deal, spending virtually the entire day feeding. They take the most nutritious parts of plants, the new growth, which has not had time to build up structural fibres. In all wildfowl species the legs are set far back on the body, making walking difficult, this being especially true of the swans, whose legs are also short; though swans are more terrestrial than, say, divers, they are essentially aquatic birds and do not stray far from water. As with other wildfowl swans moult their flight feathers simultaneously and so are flightless for a period of time which may last up to six weeks.

Swans are the largest and heaviest of the wildfowl. There are seven species, of which four (or three depending on your view of the taxonomy of tundra swans) are Arctic birds. The distribution of the species is curious; apart from the northern swans there is one in southern Eurasia, one in southern South America and one in Australia/New Zealand. The northern species are predominantly white and have very long necks. Male, female and winter plumages are similar: in this case swans mirror geese, but differ from ducks, probably because while swans and geese mate for life, ducks do not and so require conspicuous male plumage as part of mating rituals. Being large and aggressive, female swans and geese are less vulnerable to predators and so do not require the cryptic plumage of cryptic ducks (or eclipse drakes).

Swans are strong fliers once they have taken off. Take-off follows an energetic (and awesome) race across the water: Photo 53c. Landing swans are a great sight, feet splayed out as brakes and landing carriage, with a good touchdown seemingly as much of a surprise to the swan as to the onlooker. Landings on ice are hilarious, though presumably traumatic for the bird, and a great deal more traumatic for other birds already on the surface as they are skittled by the on-rushing swan. Nests are untidy heaps of material (Photos 53b and 55e) sited on, or close to, water. Swans are wholly or primarily vegetarian, grazing aquatic vegetation and dabbling in shallow water. They also up-end to feed, an efficient, if somewhat inelegant, pose. Dabbling takes in aquatic invertebrates as well as vegetation, and some swans paddle with their feet in order to bring larvae to the surface. They are, however, opportunistic and have been seen to swallow sizeable fish. Wintering birds feed in cereal feeds and on waste grain, and will grub for tubers and potatoes, the latter a relatively recent addition to the diet.

All northern species are migratory, some travelling great distances, though these are usually accomplished in relatively short flights between 'refuelling' stops. Swans have been seen at heights of more than 8,000m, but chiefly fly at 2,000–3,000m.→

Whooper Swan *Cygnus cygnus*
The plumage is white, occasionally with a pale yellow wash to the head and neck. Head and neck also sometimes tinged red from iron deposits in the feeding lake. The bill is black with a yellow base. The yellow coloration is variable, but usually extends to and beyond the nostril. As a diagnostic, Whoopers have the most extensive yellow on the bill of any of the northern swans. Whoopers are the most accomplished walkers of the northern swans, a point worth recalling if tempted to approach too closely – Whoopers are intolerant of, and aggressive towards, intruders.

Breeds in Iceland and across Arctic Eurasia to Kamchatka (though uncommonly in Norway); has also bred in western Alaska. In general a bird of the taiga, but nests on tundra in both Iceland and Russia. Some wintering birds are resident, but most move to southern rivers and coasts.

a–f Whooper Swans

Tundra Swan *Cygnus columbianus*

There are now considered to be two subspecies that in the past have been thought to be full species.

Whistling Swan *C. c. columbianus*

The plumage is white, the bill black with minimal yellow at the base. The yellow base is diagnostic in those areas of Alaska where Whistling and Trumpeter swans overlap, but to confuse the issue there are instances of Whistling Swans having all black bills. The northern habitat of the birds, which nest beside tundra pools but occasionally some distance from water, has led to the birds frequently being called Tundra Swans.

Whistling Swans breed on the Aleutian islands, western Alaska and across northern North America to the western Ungava Peninsula, including the southern Canadian Arctic islands. Absent from north-eastern Canada, but breeds in places on the southern shores of Hudson Bay; also breeds in eastern Chukotka.

Bewick's Swan *C. c. bewickii*

The plumage is white, the bill black with a yellow base. The bill colour pattern is distinctive so that individual swans can be recognised. Similar to Whistling Swan but with more yellow on the bill, though generally smaller. Interbreeding is known to occur where the two swans overlap in eastern Chukotka.

Bewick's Swans breed across Arctic Russia from the oo to western Chukotka. Swans breeding west of the Urals migrate to the British Isles, while those to the east head for Japan, Korea and south-east China.

Trumpeter Swan *Cygnus buccinator*

The plumage is white overall, occasionally rust-stained on head and upper neck from iron deposits in feeding lake. Bill is black, with no basal yellow. The scientific name of the species derives from the Latin for a military trumpet. As a family the swans are a noisy bunch (though the Mute Swan – *Cygnus olor* – while definitely not mute is less noisy), the name 'swan' apparently deriving from the Saxon for noise. The Trumpeter, which has a long, twisted windpipe, makes a deep, resonant bugling which is the loudest of any wildfowl and ranks with the loudest of all birds.

Trumpeter Swan is not truly Arctic, breeding in south-eastern Alaska, though is occasionally seen in western and central Alaska.

→ The grace and beauty of swans has inspired people for thousands of years. The Greek legends of Leda and the Swan, and of Phaeton and Cygnus (the latter story giving its name to the genus – Cygnus – as well as 'cygnet', the name of a young swan) are early examples, while a later Scandinavian tale was the basis for the ballet *Swan Lake*. For sheer elegance, a swimming swan is hard to beat. The bird is occasionally mocked, the suggestion being that the above surface elegance is belied by frantically paddling feet below the water. In reality, the paddle stoke is usually leisurely and every bit as elegant.

The grace that excited poets did not always stretch to the more pragmatic human; swans are large birds and make good eating. Northern native peoples have always prized the swan for food, and also for the luxury of its feathers; the preferred bed of the Inupiat of northern Alaska was swan skin. In medieval England swans were owned by the sovereign who was the only person allowed to harvest the birds, the penalty for poaching being severe. Even today the British monarch owns the majority of swans on the River Thames; the remainder are owned by one of two ancient Companies, the Dyers and the Vintners. The annual catching and marking, by nicks on the bill, of this year's cygnets – a ceremony called swan-upping – is typically British and would thoroughly confuse an Inuk.

---

a, c, d Whistling Swans.
b, e, f Bewick's Swans. 'e' is nesting, 'f' is a feeding family.
g, h Trumpeter Swans (and Mallards).

Geese

Geese have longer legs than their wildfowl cousins, and the legs are also set more centrally on the body. The birds are therefore more mobile, in keeping with their more terrestrial habits. Though geese can occasionally look awkward on land, they are surprisingly quick, as a traveller who blunders close to a goose nest will rapidly discover. The cryptic incubating bird can show a remarkable burst of speed which, allied to a very aggressive nature, can lead to a worrying few seconds (Photos 63b, c and d).

The 'true' geese – birds of the genera Anser/Chen (grey geese - the division of the species into two genera is disputed: some experts consider all should be grouped in Anser, but the two genera are maintained here) and Branta (black geese) – are all found in the northern hemisphere. The colour groupings are not exact, as we shall see, and are not the primary reason for genera separation. The sexes of all northern geese are alike, as are breeding and winter plumages. Twelve (of the 15 species) are Arctic dwellers. All are essentially terrestrial, though they do feed in water and may even up-end. On land they consume vegetation, while the larger species also grub for roots and tubers. In winter they feed on agricultural land, taking waste grain, but also more substantial foods such as beets and potatoes.

Geese migrate considerable distances, but they cannot afford the weight of a large digestive system that would maximise the nutrient extraction from their poor diet. They therefore take fresh green shoots, which are more easily digested and are rich in protein and carbohydrates. Migrations are timed so the geese move north in continuous spring feeding on abundant green shoots, occasionally moving to high ground to access an 'altitudinal' spring.

Grey geese differ from their black cousins not only in colour, but also in having serrated mandibles and vertical furrowing of the neck feathers. The latter is prominent, particularly when the feathers are vibrated, a sign of aggression. →

Snow Goose *Chen caerulescens*
Unmistakable, but seen in two colour morphs, white and blue. The white morph is entirely white apart from black primary feathers and pale grey coverts. The head and upper neck are occasionally tinged pale yellow and may also be stained red by iron in feeding ponds. The bill is red with a white nail and has prominent dark edges, giving the bird a pronounced 'grin' that cannot be mistaken (Photo 57d). The 'blue goose', once thought to be a separate species, has blue-grey lower neck, breast, belly, flanks and mantle. Intermediate forms also exist with many flocks including a range of colours. Greenlandic birds are considered a subspecies, as is the 'Greater Snow Goose' of Baffin, Bylot and Ellesmere Islands.

The Snow Goose is essentially a Nearctic species, breeding across North America and in north-west Greenland. However, a small population also breeds in north-east Russia, with a more significant population on Wrangel Island. In winter, North American birds fly south to the southern states, from California to Florida, and to Mexico. Greenland birds join them, as do many Russian birds, though some of the latter head south to China and Japan.

Ross's Goose *Chen rossii*
Plumage as the Snow Goose, though blue morphs are much rarer. Some authorities believe that blue morphs result from hybridisation between Ross's Goose and blue-morph Snow Goose (perhaps following egg-dumping by a blue goose into a Ross's nest, with the chicks imprinting on its foster parent). The bill of Ross's Goose differs from that of its larger cousin, having no, or minimal, 'grin' and blue-grey 'caruncles' around the base (Photo 57e).

Ross's Goose breeds in just a few places in Arctic Canada – Banks and Southampton Islands and the mainland near Bathurst Inlet, and the western shores of Hudson Bay. Wintering birds were originally seen only in California, but in recent times the geese have extended this range eastwards to the mid-continent, though the wintering range is not continuous, several isolated populations being observed.

a, d Snow Geese.
b, c Mixed flocks of white and blue Snow Geese.
e, f Ross's Geese.

Greater White-fronted Goose *Anser albifrons*

Plumage similar to Greylag Goose, but brown rather than grey-brown, the head and neck distinctly brown. The underparts are pale brown, darker on the flanks, with dark brown/black horizontal stripes. The vent and undertail are white. The bill is pink, with a large white basal ring. The legs and feet are orange. Five subspecies are recognised across the range, which is the broadest of any northern goose, and the only one to occur (in significant numbers) in both the Palearctic and Nearctic.

White-fronted Goose breeds in Russia eastwards from the White Sea, including the southern island of Novaya Zemlya, in west Greenland around Disko Bay, Alaska and in isolated areas of northern Canada. In North America migrating birds head south to California and Mexico. Birds from Asian Russia winter in China and Japan, European Russian birds head for the British Isles and Low Countries. Greenland White-fronts fly to Iceland, but this is only a staging post on the way to Ireland and Scotland.

Lesser White-fronted Goose *Anser erythropus*

Status: IUCN: Vulnerable (ongoing, rapid population decline as a result of hunting on migration route and wintering grounds, and habitat loss).

Similar to Greater White-fronted Goose but smaller with a shorter neck. The bill is pink with a white nail. The basal ring is white and extends to the crown, and there is a distinct yellow orbital ring. These features are important where Lesser and Greater White-fronted Geese overlap in Russia. One subspecies of the Greater White-fronted also sometimes shows a yellow ring. However, this is the so-called Tule Goose (*A. a. gambeli*) which is found only near Alaska's Cook Inlet where Lesser White-fronteds do not occur.

Lesser White-fronted Geese breed in northern Fennoscandia, where they have been reintroduced following severe depletion due to overhunting. However, hunting is still a problem, as is habitat loss at wintering sites. The population is probably no more than 30,000 and decreasing. Wintering birds are seen across northern Europe from east England (where it is a very rare vagrant) to eastern Germany, in the Balkans and near the Caspian Sea, and in parts of China.

→Though it was long assumed that females laid as soon as they reached their nest sites, it is now known that many geese actually spend the first few days feeding continuously to replenish bodily reserves lost during the flight north. Egg-laying cannot be delayed for long, however, as the best strategy for successfully raising young is to have them hatch early. The need to be in good condition for laying while laying early represents a dilemma for the female goose. The female grows many ovarian follicles, but will choose how many eggs she will produce. In bad years, when poor weather means the migration flight has been poorly timed and food resources are limited, the female may resorb the follicles and lay no eggs. Some females lay too many eggs. If the female cannot find enough food to continue incubation she has two choices, and both are known to be taken: some females desert the eggs, saving herself in order to breed the following year, but some have been known to die of starvation on the nest. Geese are gregarious in winter, but rather less tolerant when breeding, though colonial breeding is not rare.

Even if incubation is successful, chick-rearing is stressful and fraught with danger. There are foxes, skuas, Ravens and other predators, both avian and mammalian, to whom goslings are an easy and welcome meal. The climate, too, plays a part – a spell of bad weather may kill off the vulnerable chicks, or a poor growing season for plants may not allow some chicks to fatten sufficiently to make it to winter quarters. To increase the likelihood of successful chick rearing, both parent birds brood and guard their offspring and at colonial nest sites crèches of older chicks may also form, guarded by many adults. Life for northern geese is harsh, and that is without the extra problems caused by human hunters.

a, b, c, e Greater White-fronted Geese.
d Lesser White-fronted Goose.
f Mixed flock of Bean, Greylag and Greater White-fronted Geese.

Bean Goose *Anser fabalis*

Very similar to White-fronted Goose but lacking the dark striping on the underparts, and the upperparts are greyer. The bill is yellowish orange, usually with a dark grey base and dark grey tip or nail. Some birds show a narrow basal white ring. There are two distinct forms, which are now sometimes considered to be separate species. The 'Tundra' Bean Goose (*A. f. rossicus*) is as described. The 'Taiga' Bean Goose (*A. f. fabalis*), the larger of the two, has a shorter, thicker bill which is black, with a yellow band close to the nail. Several subspecies are also identified, Tundra Beans having three, these depending on size, colour of the belly and the form of the bill. Taiga Beans have two forms separated by size and bill thickness.

Taiga Bean Geese breed in Scandinavia east to Lake Baikal. Tundra Bean Geese breed on the northern Russian tundra east of the Urals and on the southern island of Novaya Zemlya. In winter, western birds move to southern Scandinavia, Germany and Belgium, with some being seen in the British Isles. Eastern birds move to China and Japan.

Emperor Goose *Chen canagicus*

A small, exquisitely marked goose, the only grey goose to show an exotic plumage. The head and back of the neck are white, the front of the neck dark grey. The rest of the bird is silver-grey, flecked and barred with dark grey and white. The tail is white. The short bill is dull pink. Emperors were named for the Russian Tsar (the second part of the scientific name refers to Kanaga Island in the Aleutians where they were first observed by Russian explorers/fur hunters). They are the most musical of the grey geese, with a bisyllabic (occasionally trisyllabic) call . They are also the most maritime, though rarely seen far from the coast.

Emperor Geese breed along the Aleutian chain and on the south-west Alaskan mainland, and, in small numbers, on the east coast of Chukotka. Many birds are resident, but some move to southerly Aleutians and Kodiak Island and, occasionally, further south along the western seaboard of the United States.

Pink-footed Goose *Anser brachyrhynchus*

The head and neck are brown, the rest of the bird grey or grey-brown, the underparts paler except on the flanks. The vent and undertail are white. The legs and feet are, as might be expected, pink. The colour of the feet is a useful diagnostic – the feet of Greylags are also pink, but they breed much further south. The bill is the best diagnostic characteristic when confronted with swimming birds; it is short and brown, with a pink band behind the brown nail.

Pink-footed Geese breed in east Greenland, mainly north from Scoresbysund, on Iceland, chiefly in the interior, though in isolated areas on the coast as well, and on Svalbard. Pairs have been noted on Franz Josef Land, but there is no evidence of successful breeding. In winter western birds head to northern England and Scotland, while Svalbard birds fly to Norway, Denmark and Holland.

Greylag Goose *Anser anser*

The largest grey goose, only the largest Bean Geese approach it in size. Greylags also have the 'honk' which is associated with all geese, not surprisingly as it is the ancestor of farmyard geese. Overall grey-brown, the chest and belly paler, sometimes with darker striping. The vent and undertail are white. The bill is pink. The legs and feet are dull pink (though often stated, the name is not a mis-spelling of 'grey leg' but seems to derive from the fact that the species migrated later than, i.e. 'lagged' other goose species. The neck furrowing of Greylags is very pronounced, particularly when a bird charges a real or imaginary intruder prior to a 'triumph' display, this involving considerable loud honking.

Greylag Geese breed in Iceland and northern Norway, otherwise only in more southerly areas of Eurasia. The species is not, therefore, a true Arctic dweller. Wintering birds fly as far as northern India and southern China.

a,b Bean Geese.
c, d Emperor Geese.
e Pink-footed Geese.
f, g Greylag Geese.

While the black geese are generally darker than the grey, the distinguishing features of the five northern species are the bold patterning – only the Emperor Goose breaks the rather dull monotony of the greys – and the lack of prominent mandible serrations. Of the five only one, the Hawaiian Goose (or Nene, *Branta sandvicensis*), shows any furrowing of the neck feathers – Nenes are not an Arctic species.

Canada Goose

It has always been clear that there were two distinct groupings of Canada Geese, these being called, at one time, the Greater and Lesser Canadas because of the size differential between those birds usually seen on the tundra and those from more southerly latitudes. However, in all cases the birds were considered to be subspecies of *Branta canadensis*. The taxonomic status of the Canada Geese has recently changed, mtDNA studies having identified the existence of two matriarchal clades which correspond to the large and small bodied Canadas, indicating that they separated about 1 million years ago. The American Ornithological Union has therefore split the Canada Goose from the Cackling Goose. While the change is sound genetically, it is not the case that all Cackling Goose subspecies are smaller than all Canada Goose subspecies, nor that the Cackling Goose is the more northerly species. 'Cackling' also seems a somewhat unfortunate name for a beautiful bird.

Canada Goose *Branta canadensis*

Seven subspecies nesting south of the tundra. Head and neck black apart from a prominent white band which extends from the throat to the ear. The upperparts are pale grey-brown, the underparts paler, but with darker flecking on the flanks. The vent is white, the tail black. The subspecies include the largest black goose (the Giant Canada Goose *B. c. maxima*, which breeds in south-west Canada/north-west USA).

Canada Geese breed across North America. However, as they are attractive birds they have introduced into, or escaped captivity in, other countries, including the British Isles, Scandinavia and other areas of northern Europe. All northern birds winter in the southern USA.

Cackling Goose *Branta hutchinsii*

Status: IUCN: Least Concern (though *B. h. leucopareia* not specifically assessed).
CITES: *B. h. leucopareia is Appendix I.*
Four subspecies which are identical in pattern to the Canada Goose, the differences between the two being related to size alone, though there are some subtle differences - for instance the nominate, also occasionally called Richardson's Goose, of NWT and Nunavut, has a marginally different head shape and a stubby bill in comparison to Canada Geese (Photo 63f). The smallest of the subspecies is *B. h. minima* of north-west Alaska. The Aleutian Canada Goose (*B. h. leucopareia*) once bred on all the islands of the Aleutian chain. To increase fur production, Russian settlers introduced Arctic Foxes to all the readily accessible islands. The effect on the geese, and other ground nesting birds, was catastrophic, and it was feared they had become extinct. But they were found breeding on inaccessible islands, and a programme of fox elimination and goose re-introductions has ensured the survival of the subspecies. Sadly a fifth subspecies, *B. h. asiatica*, which bred on the Kuril and Commander islands, is now thought to be extinct as no specimen has been veritably sighted since 1914.

Cackling Geese breed across the North American tundra, on some Aleutian Islands. There is a small population in west Greenland. All birds winter in the southern USA.

a–d *Canada Geese (*B. c. interior*).*
e *Cackling Goose (*B. c. taverneri*).*
f–h *Cackling Goose (*B. h. hutchinsii*).*
i *Mixed spring flock of Cackling Geese and Snow Geese, Barren Lands, NWT.*

Barnacle Goose *Branta leucopsis*

A beautiful small goose. The face is white, apart from a black stripe from bill to eye. The rest of the head and neck is black. The upperparts are white and pale grey, with black barring. The underparts are white with pale grey barring. The vent and rump are white, the tail black. Each winter the geese appeared in northern Europe, but as the origin of the birds was unknown they puzzled locals, and a belief arose that they hatched from the barnacles attached to flotsam wood that occasionally washed ashore. Writing in the early 17th century, the herbalist Gerard noted an extension of the legend – *'There are found in the North parts of Scotland and the Islands adjacent, called Orchades, certain trees whereon do grow certaine shells of a white colour tending to russet, wherein are contained little living creatures: which shells in time of maturitie do open, and out of them grow those little living things, which falling into the water do become fowles, which we call Barnacle Geese ... and in Lancashire, Tree Geese: but the other that fall upon the land perish and come to nothing'.* So the myth had extended, the fact that barnacles washed ashore were invariably attached to wood meaning that there must, somewhere, be a tree with very strange fruit. Eventually, of course, the truth was discovered, but the name (Barnacle rather than Tree) stuck, in part because barnacles, being 'fish' could be eaten on Fridays by Christians and the goose, a 'fish goose', could also be eaten as fish.

Barnacle Geese have a restricted range, breeding only in north-east Greenland, Svalbard, Novaya Zemlya's southern island and nearby parts of the Russian mainland. In recent years Barnacle Geese have also bred on Iceland. Greenland and Svalbard birds winter in Scotland and Ireland, the Russian birds in Denmark, Germany and the Low Countries.

Brant Geese (occasionally Brent Geese) *Branta bernicla*

Black head and neck with a triangular black-and-white striped patch on each side of the neck. Remaining plumage is dark grey, the underparts paler (though often not much paler). The vent and undertail are white.

Brant Geese breed in northern Greenland, Svalbard, Franz Josef Land, across Arctic Russia east from the Taimyr Peninsula, including Severnaya Zemlya, the New Siberian Islands and Wrangel Island, and across northern North America from Alaska to western Hudson Bay and on all Canada's Arctic islands. The nominate race (as described), often called the Dark-bellied Brant, breeds in Arctic Russia apart from Franz Josef Land. There, and on Svalbard, Greenland and eastern Canada, the Light-bellied Brant (*B. b. hrota*) breeds. It has paler underparts with darker flecking on the flanks. The Black Brant (*B. b. nigricans*) breeds in western North America. It is usually much darker overall. However, this tidy organisation is confounded by some Black Brants having underparts as pale as nominate, by the fact that some authorities claim that in Russia, apart from on the Taimyr Peninsula and Franz Josef land, all breeding birds are Black Brants, by the existence of intermediate forms, and by the general distribution of colouring in all forms.

In winter birds, from Greenland, Svalbard, Franz Josef and western Russia fly to the Low Countries and British Isles. Those from eastern Russia move to Japan, while American birds winter on both coasts of the United States.

Red-breasted Goose *Branta ruficollis*

Status: IUCN: Vulnerable (as population is not well-understood). CITES: Appendix II.
The most attractive black goose. The head and neck are patterned black, white and red-brown, the red-brown of the neck extending to the breast. The upperparts are black with white stripes on the coverts, the underparts black with white barring on the flanks.

Red-breasted Geese formerly bred only on Russia's Taimyr Peninsula, but in recent years have extended their range both east and west. Despite this good news, the position of the goose remains vulnerable. Wintering birds are concentrated in a small number of places on the Black Sea where changes to local agricultural methods represent a threat. The birds are also hunted as they migrate.

a–c Barnacle Geese.
*d Brant Goose (*B. b. hrota*) photographed on migration through Iceland.*
*e Brant Goose (*B. b. nigrans*).*
f Red-breasted Goose and goslings.

Ducks

Dabbling Ducks

Dabbling ducks are the largest duck group, a highly successful group with representatives on all continents except Antarctica (though some species breed on Southern Ocean islands). These ducks are named for their habit of working the surface of the water for food. They feed chiefly in fresh water, though they are not unknown in marine areas. They primarily feed on aquatic vegetation (but not fish), though will also take aquatic invertebrates. Dabbling involves taking in a volume of water that is then squeezed out through comb-like lamellae (the pecten) at the edges of the mandibles, seeds and other particles then being swallowed. This feeding method is analogous to that of the large whales (and as such can be considered an example of convergent evolution). The ducks often feed by upending (Photo 67a).

The various species have long, broad wings which allow not only a fast flight but also a short take-off, some ducks being able to rise almost vertically from the water if necessary. The ducks are sexually dimorphic, males (drakes) usually being brightly coloured during the breeding season, while the females (ducks) are cryptic brown as camouflage. As noted previously, it is probable that the colourful male plumage is related to the fact that ducks pair annually rather than for life, the male needing fancy plumage to attract females. But the plumage would attract predators during the flightless moult, so after the breeding season the males adopt a more cryptic 'eclipse' plumage. Female plumage is cryptic throughout the year. Drakes have a coloured speculum, which is maintained in eclipse. Females usually have a speculum as well, but it may be much smaller or less clear-cut. →

Mallard   *Anas platyrhynchos*
The most widespread and probably most recognisable of the dabbling ducks. Drakes are handsome birds with metallic green sheen on heads and necks. The speculum is blue, bordered by thin black and wider white bands. The female is cryptic brown but with the same speculum. Eclipse drakes are as female, but darker. Drake Mallards are one of few birds with a penis and often engage in what can only be termed rape, the female being held down by clasping the nape of the neck, often causing feather loss and skin damage (Photo 67b).

Mallards breed in west and east Greenland (these are the most Arctic of all Mallards and are considered a subspecies, *A. p. conboschas*), in Iceland, and across Eurasia (though only to the northern coast in Fennoscandia and then increasingly southerly to the east, though occurs in Kamchatka). In North America, Mallards breed throughout Alaska to the Mackenzie delta, but more southerly across Canada. Greenland and Icelandic birds are resident, but northern birds in Eurasia and the North America move south for the winter.

American Black Duck   *Anas rubripes*
Somewhat poorly named as the drake is not black, having a brown head with darker crown and eye stripe. The rest of the bird is dark brown. The speculum is purple, bordered by dark brown bands. Female is cryptic brown, with the same speculum. Eclipse drake is as breeding but washed grey

Black Ducks have been severely overhunted, with consequent population reduction, and now faces two further threats. One is the loss of its preferred habitat, hardwood wetlands. The other is more serious, the northern advance of the Mallard, a highly adaptable species which tends to outcompete the Black Duck, and also to hybridise with them, pure-bred Black Ducks becoming rarer. Black Ducks are more cold-resistant, but as the Earth warms and Mallards adapt, this is likely to become an increasingly less useful advantage.

Black Ducks breed around the southern shores of Hudson Bay and northward into Quebec and Labrador, though rarely to the northern coast.

a-f Mallard.
g American Black Duck (male).
h American Black Duck (female).

American Wigeon *Anas americana*

Breeding drakes have black-speckled buff heads with a cream forehead and green crescent to the rear of the eye. The upperparts are brown-speckled grey, the underparts salmon pink. The speculum is black with a central band of green. Females are cryptic rufous brown or dark brown with a similar speculum but reduced white patch. Eclipse drakes are as female, but brighter.

American Wigeons breed across northern North America, but only within the Arctic in western Alaska, near the Mackenzie delta and southern Hudson Bay. Wintering birds are seen in the southern United States.

Eurasian Wigeon *Anas penelope*

As elegant as their North American cousins, with which they will hybridise. Breeding drakes have dark brown heads with buff forehead. The upperparts are as the American Wigeon, but the breast is pink-buff, the belly white with pale grey flanks. The speculum is dark grey bordered by black bands. Female is as American Wigeon, but darker. Eclipse drake is as female, but more with more rufous underparts and brighter overall.

Eurasian Wigeon breeds in Iceland and across Eurasia, though not to the northern coast except in Fennoscandia and European Russia. Birds from northern Iceland migrate to the south of the island, though some move to the east coast of North America. Eurasian birds head south to Japan and central Asia.

Baikal Teal *Anas formosa*

Status: IUCN: Least Concern (recently upgraded from Vulnerable as population now increasing). CITES: Appendix II

Arguably the most exquisite of all the northern ducks; drakes have marvellously patterned heads, the crown dark grey, the face buff-yellow, split in two by a narrow black line from the eye to the black chin. The nape is bottle-green, this continuing towards the buff-yellow throat as a thinning curve. The breast is salmon pink merging with a white belly and pale steel-grey flanks. The upperparts are streaked brown, grey and white. The speculum is green with a rufous band at the front and broad white band at the rear. Female is cryptic brown, but with cream underparts and a distinct white loral spot. Eclipse drake as female but more rufous overall.

Baikal Teal breeds from Russia's Taimyr Peninsula east to Chukotka, but patchily. Also breeds east from Lake Baikal, around the shores of the Sea of Okhotsk and in Kamchatka. Mainly resident, but northern birds move to Korea and eastern China. Formerly the most numerous of Asian ducks, but decimated by hunting. The habit of feeding in arable fields in winter has led to large-scale killing with poisoned grain.

→Pair bonding is seasonal, often occurring during the winter or spring migration. Male ducks have limited displays, the female invariably unmoved by the performance and rarely joining in. The Mallard drake's attitude to mating has already been noted (previous page, species description), but the drakes of other dabbling ducks also forcibly inseminate available females; in some circumstances several drakes will attempt to forcibly mate with a female (sometimes the female is forced under the water by the relentless activity of squabbling drakes and drowns).

In general, insemination is the only contribution of the drake, with the males deserting the females after they have laid. Ducklings are precocial, able to swim and feed almost immediately upon hatching, but they are brooded at first by the female, and cared for until fledged. Females are diligent in their care of their young.

a American Wigeon pair.
b Male American Wigeon.
c Male Eurasian Wigeon.
d, e Male Baikal Teal.
f Eurasian Wigeons.

Northern Shoveler *Anas clyptea*

The spatulate bill of this species is diagnostic. Drakes are handsome birds, with a dark green sheen on the head and neck, and mottled grey and white upperparts. The breast is white, the belly and flanks chestnut. The speculum is green with a white band at the front. Female is cryptic brown and white, with a duller speculum. Eclipse drake is as female, but brighter and with a greyer head and neck. Male shovelers differ from other dabbling ducks in being less promiscuous, though their faithfulness to their mates does not extend to staying long after incubation starts, often abandoning the female the moment she starts to brood her clutch.

Shovelers breed in Iceland and across Eurasia, but only to the Arctic fringe. In North America they are confined to the area west of Hudson Bay, though they can also be seen on the southern shores of the bay. Also breeds to the Mackenzie delta, though more southerly in Alaska.

Northern Pintail *Anas acuta*

With their elongated tail feathers, drake Pintails are both extremely handsome and unmistakable. They are elegantly patterned, with a dark brown head and upper neck. The front of the neck, breast and belly are white, the flanks speckled grey. The upperparts are grey and black. The speculum is metallic dark green bordered by a buff band at the front and a white band at the rear. Females are cryptic brown and cream. They also have elongated tail feathers, though these are not as long as those of the drake. The speculum is dark brown, bordered front and rear by white bands. Eclipse drakes are as female, but paler and with longer tail feathers. Drakes will defend their mates against mating attacks by other males, but are also willing to involve themselves in such attacks elsewhere.

Northern Pintails breed across northern Eurasia, mostly to the coast, but not on any Arctic islands. Breeds in Iceland, but absent from Greenland. Breeds across North America as far east as the Ungava Peninsula, to the north coast except on the Boothia Peninsula, and also on southern Banks Island. Wintering birds are found in central Africa, central Asia, Japan, the southern United States, Mexico and Central America.

Eurasian Teal *Anas crecca*

The smallest of the northern Anas ducks. Drake has a chestnut head, the coloration interrupted by an upside-down horizontal 'comma' of dark green from the eye to the nape. The comma is delineated by narrow buffish-yellow bands. The upperparts are grey and grey-brown. The breast is grey-speckled buff, the rest of the underparts grey with darker wavy barring. The speculum is metallic green and black, with white/buff bands at the front and rear. Female is cryptic brown, with a similar speculum enclosed in narrower bands. Eclipse drake is as female, but darker.

Eurasian Teals breed in Iceland and across Eurasia, but rarely to the north coast. They fly to Japan, southern Asia and the Middle East for the winter.

Green-winged Teal *Anas carolinensis*

Until recently considered a subspecies of the Eurasian Teal but now accorded species status as a consequence of *cytb* and mtDNA sequencing. Drake is as Eurasian Teal but lacks the delineating bands around the head comma and shows much less white on the speculum bars. Females and eclipse drake are as Eurasian Teal.

Breeds across North America, including west and north Alaska and northern Yukon/ western North-West Territories. Breeds around southern Hudson Bay, near Ungava Bay and in northern Labrador. The birds of the Aleutian Islands are considered a subspecies, *A. c. nimia*: drakes of these resident teals are larger than those of the nominate, but otherwise identical.

> *a Male Northern Shoveler (and male Mallard).*
> *b Male Northern Shoveler.*
> *c Male Northern Pintail.*
> *d Female Eurasian Teal.*
> *e Male Eurasian Teal.*
> *f Female Green-winged Teal.*
> *g Male Green-winged Teal.*

Pochards

Pochards, the members of the genus Aythya, share the same diet as dabbling ducks, but primarily feed with shallow dives, usually to depths of a few metres. Pochards are heavier than dabblers, they have longer necks, and the feet are set further back on the body to act as more efficient paddles. This makes them awkward on land and they are rarely seen far away from water, or indeed, out of the water at all. Females make nests on floating mats of vegetation or in reed/sedge beds, only rarely at the water's edge. Their heavier weight means pochards have more difficulty becoming airborne, with take-off following a run across the water, a sharp contrast to the often near-vertical lift-off of dabblers.

In general, pochards are drabber than dabbling ducks, and lack the colourful speculum. However, they share some behavioural habits; for example, the drakes desert the females after they have begun incubation. The males then tend to congregate for the moult. Females stay with their young until they begin the moult at which point the ducklings often form large crèches until they fledge.

Pochard sometimes interbreed. In this respect they are not unlike some of the dabbling species and also some goose species. However, pochards do so much more readily, a fact that can cause the observer problems. Why ducks and geese in general, and pochard in particular, are apt to interbreed is not understood.

Tufted Duck *Aythya fuligula*

Drake has a black head with a distinct purple sheen and a long, downcurved crest. The upperparts are black, the underparts white. Female has a dark brown head and upperparts, the breast lighter brown, the belly white. Females show a smaller crest, but have the same white wing-stripe as the drakes. Eclipse drakes are as breeding but much duller on the head and underparts, and the white underparts are heavily flecked brown. Found in fresh and brackish waters were they are omnivorous.

Tufted Ducks breed in Iceland across Eurasia, though only to the northern coast in Fennoscandia. The ducks breeds on Kamchatka. Wintering birds are largely resident but some move to the Mediterranean coasts of southern Europe and north Africa, the coasts of the Black and Caspian seas, across central Asia and the coasts of Japan, Korea and south-east China.

Canvasback *Aythya valisineria*

Drake Canvasbacks have handsome rufous heads and necks, and a black breast. The upperparts are whitish grey (this feature giving the bird its name), as are the underparts apart from the undertail-coverts, which are black. Female has the same pattern, but the head, neck and breast are brown with white flecking, and the upper- and underparts are brown, speckled pale grey. Eclipse drake is as breeding but head, neck and breast are dark brown, with the rest of the plumage as the female. Usually feed by diving, taking both vegetation and benthic invertebrates.

Canvasbacks breed in a band across central Alaska from the west coast, but are scarce throughout much of the range. The ducks breed to the Mackenzie delta, but eastwards are found much more southerly as far as the Ontario border. The species is absent from eastern Canada. In winter the birds fly to areas of the east and west coasts of the United States, the Gulf of Mexico and the Mississippi delta.

In central Alaska the birds may be confused with the Redhead (*Aythya americana*). The latter has a redder head, a less heavy (and bluer) bill, and greyer upper- and underparts.

a *Female Tufted Duck and brood.*
b *Female Tufted Duck.*
c *Male Tufted Duck.*
d *Male Canvasback.*
e *Female Canvasback.*

Lesser Scaup *Aythya affinis*

Drake has a black head with a purple gloss (not always easily distinguishable). The neck and breast are black, the upperparts white, vermiculated dark grey and black. The underparts are white with grey flecks on the flanks. Female is overall grey-brown. There is a distinct white basal ring to the bill. Eclipse drake as breeding but much duller, with the upperparts more grey-brown and the underparts flecked grey-brown. The patterning is also much less well-defined. Found in fresh and brackish waters, the latter if they are only moderately salty. Feeds mostly by diving, with a diet comprising chiefly of animal matter – insects, crustaceans and molluscs – but also including aquatic plants.

Lesser Scaup breed across North America, though to the north coast only near the Mackenzie delta. Also breeds on part of the western shore and most of the south-eastern shore of Hudson Bay, but otherwise subarctic. Wintering birds are seen along the east and west coasts of the United States and Mexico.

Greater Scaup *Aythya marila*

Very similar to Lesser Scaup, but the head gloss is green rather than purple and the upperparts tend to be lighter. Female and eclipse drake are also similar to Lesser Scaup. In the Nearctic where both species breed there is ample scope for confusion. In general the head gloss is diagnostic, but in some lights the sheen colours can be reversed, which can render the difference useless for identification purposes. Head shape is usually a more reliable characteristic. The Lesser Scaup has a tall, narrow head, whereas the Greater Scaup's head is wider and more oval. Head shape when viewed from the side also differs, the Lesser Scaup's head being squarer, the crown and nape flatter, while the Greater Scaup's head is more rounded and has a peak on the crown. Caution is still required as the head shape can change with activity. In the Palearctic, where the Lesser Scaup does not breed, there is confusion with Tufted Duck; the crest of the Tufted drake and the more clearly discernible purple head gloss are diagnostic. In females the crest of the Tufted Duck and the white facial ring at the base of the bill of the Scaup female are diagnostic. Favours freshwater ponds and lakes, and occasionally found on rivers. Dabbles and upends as well as feeding by diving, considering which it is no surprise the diet is broad, with animal and vegetable matter.

Greater Scaup breed in Iceland and across Eurasia, though only to the coast in Fennoscandia and Russia to the Taimyr Peninsula. Also breeds on Kamchatka. Nearctic wintering birds share the same range as the Lesser Scaup. Western Palearctic birds head to the coasts of southern Scandinavia, the British Isles, the Low Countries and France, and the Mediterranean, Black and Caspian seas. Eastern Palearctic birds move to the coasts of Japan, Korea and China.

a, b Lesser Scaup.
c Female Greater Scaup.
d Male Greater Scaup.
e Greater Scaup.

Eiders

The four eiders are the most maritime of all wildfowl, and are pelagic for much of their lives. The eiders are dive-feeders, making the deepest dives of any duck. Common Eiders are known to dive to 20m at least and it is believed that the Spectacled Eider may reach depths of 50m. However, eiders usually dive to no more than 5m. During dives the birds use their feet as paddles, with occasional wing strokes. The prey comprises molluscs, which are prised from rocks, crustaceans and other benthic animals. Steller's Eider, the smallest of the group, feeds extensively on aquatic larvae, especially in freshwater, and its dives rarely exceed depths of c.8m. As an adaptation for diving the birds are heavy, take-off requiring a lengthy run across the surface. They are strong fliers but not manoeuvrable, a fact that is particularly noticeable when they land; landings have none of the grace of the smaller ducks, involving more of the crash-landing technique of the swans.

Drake eiders are finely patterned, the females having the familiar cryptic brown plumage required of tight-sitting nesters. The drakes parade their plumage in mating displays while calling a musical three-syllable coo. When heard through an opaque Arctic sea mist, the call is ethereal and evocative. Female eiders have an additional protective technique when incubating; if forced to flee they will, just before departing, defecate evil-smelling faeces on their eggs. This may well deter Arctic Foxes but it is of little benefit against gulls and skuas, which have a limited sense of smell. Eider ducklings frequently form large crèches of up to 100 birds (500+ have been seen) in the care of one or more females (Photo 77d).

Common Eider *Somateria mollissima*

Status: IUCN: Near threatened (population decline in Europe resulting from overhunting). Drakes have a wonderfully patterned head, something shared with males of the other two Somateria eiders. The female is cryptic brown with a brown 'speculum'. Technically only Steller's Eider and Harlequin Duck of the sea ducks have a true speculum, though several other species have colour patches on the secondaries that give a similar appearance. In the drake Common Eider this is black. Eclipse drakes have a grey-brown head, the white of breeding replaced by grey-brown. Dive for crustaceans, echinoderms and molluscs. Nests are lined with down plucked from the female's breast to insulate their eggs (Photo 77e). The value of eider down to humans in keeping the northern winter at bay has been obvious since at least the 7th century when St Cuthbert, the first Bishop of Lindisfarne, set up a sanctuary for Common Eiders on one of the Farne islands off the UK's Northumberland coast; eiders are still known by locals as Cuddy Ducks in his memory. Commercial farming was begun by the Vikings two centuries or so later, and in the early 20th century down remained a major export of Iceland. The down was collected from nests after the chicks had departed. Each nest produces about 15g of raw down, the cleaning process reducing this to about 1.5g of usable material. A kilogram of exported down therefore required the input from 700 nests; at the industry's height, Iceland exported over 4 tonnes of raw down annually, representing the output from almost 300,000 nests. The industry is now much reduced. The scientific names of the bird reflects the usage of the down. Somateria, the genus name for the three large eiders, derives from the Greek for 'down body' while *mollissima*, the specific name for the Common Eider, derives from the Latin for 'softest'.

Common Eiders breed on both coasts of Greenland (though not to the far north), on Iceland, Svalbard, Franz Josef Land, in Fennoscandia and the southern island of Novaya Zemlya. In Arctic Russia breeding is patchy, occurring on the New Siberian Islands and Wrangel Island, with isolated breeding sites on the mainland (more concentrated in Chukotka). In North America the birds breed on the Aleutians, the west and north coasts of Alaska, on the northern Canadian mainland coast (including Hudson Bay), and on the southern Arctic islands. Greenland birds winter in Iceland, where the ducks are resident. The birds of the Eurasian Arctic islands move to the southern Scandinavian coasts or the Bering Sea. Nearctic birds move to the both coasts of the United States.

a Common Eiders (and male Goosander).
b, f Male Common Eiders.
c, d Female Common Eiders. In 'd' several females are guarding a crèche of ducklings.
e Common Eider nest.

King Eider *Somateria spectabilis*

Drakes justify their name with a regal head pattern that includes an orange and black forehead shield (which the Inuit sometimes bite off and eat immediately after killing a bird). The crown and nape are pale blue, while there is a pale green patch below the eye merging to white cheeks. The upper mantle is white, the rest of the upperparts black with two 'shark's fins' arising from the tertials. The breast is cream tinged pink, the rest of the underparts black apart from white patches on the rear flanks. Females are cryptic with a dark brown 'speculum'. Eclipse drakes maintain the forehead shield, but the rest of the head is grey-brown, as is the breast. The rest of the plumage is as breeding, but duller.

Breeds in west and north-west Greenland, on the island's east coast (around Scoresbysund) and on the north-east coast. Also breeds on Svalbard, Novaya Zemlya and along the north Russian coast from the White Sea to Chukotka, including the New Siberian Islands. In North America breeds along the north coast from Alaska to Hudson Bay and on all Canada's Arctic islands. Absent from Canada's coast east of Hudson Bay. In winter some east Greenland birds fly to Iceland. West Greenland birds make for the open water off the south-west coast. Svalbard and west Russian birds move to the Barents and Kara Seas, with east Russian birds heading to the Bering Sea. American birds move to the Bering Sea and Labrador coast.

Spectacled Eider *Somateria fischeri*

The smallest and least-known of the three Somateria eiders. Drake has spectacular goggles (the spectacles of the name), the 'lens' being white, the 'frame' black. The forehead is pale green, as is the nape. The rest of the neck and upperparts are white. The underparts and tail are dark grey. Females are cryptic brown, but also have 'goggles' with pale brown lenses, the rest of the head darker. Eclipse males are as female but darker. Chicks also have 'goggles'.

Spectacled Eider breeds on the Asian Russian coast east of the Lena delta and on the New Siberian Islands and Wrangel Island. Also breeds on the western and northern coasts of Alaska. The Alaskan population crashed in the 1990s for reasons unknown and has continued to decline; the birds are now rare and much sought-after by Arctic travellers. Until the mid-1990s the wintering range of Spectacled Eiders was completely unknown. Only with the advent of radio-tracking did it become possible to follow migrating birds. This led researchers to the central Bering Sea, where photography revealed perhaps thirty flocks comprising a total of at least 150,000 birds. It is now known that the birds winter in large numbers in areas of open water, chiefly near the islands of St Lawrence and St Matthew. Aerial photography suggests that the birds help keep leads free of ice by diving and swimming.

Steller's Eider *Polysticta stelleri*

Status: IUCN: Vulnerable (rapid population decline for currently unknown reasons).
The smallest of the four eiders. The drake is the least spectacular eider, with a white head, black eye ring, pale green patches on the lores and forehead, and a curious green tuft on the back of the head. The throat, mantle, rump and tail are black. The wings are white with black primaries and a black speculum, bordered white. The underparts are buff-orange, the belly darker, but the flanks white. Females are cryptic brown with a blue speculum, bordered white. Female also has drooping tertials. Eclipse drakes are as female, but with a paler head and white flanks.

Steller's Eiders breed on the Asian Russian coast east from Khatanga Bay, and on the New Siberian Islands. May also breed on other, isolated sections of the Russian coast, perhaps as far west as the Kola Peninsula; birds have been seen off the Fennoscandia coast and in the Baltic Sea in winter. Also winters in the Bering Sea. In North America the birds breed in western and northern Alaska.

a *Male King Eider.*
b *Female King Eider.*
c *Mating King Eiders.*
d *Flock of King Eiders above the sea ice of northern Baffin Bay.*
e *Male Steller's Eider and male Common Eiders*
f *Incubating female Steller's Eider.*
g *Female Spectacled Eider.*
h *Male Spectacled Eider.*

Sea Ducks

Seaducks are a group of pelagic waterfowl (which, technically, should include the eiders, though they have been considered separately here as their flamboyant plumage makes them different from the other, essentially black-and-white, birds). Seaducks dive much deeper than the diving ducks and occasionally pursue prey, particularly the sawbills, which take fish and swimming invertebrates, while the rest chiefly feed on benthic animals. As with the eiders, shellfish and crustaceans are often swallowed whole and crushed by the muscular gizzard.

As in the dabbling and diving ducks, drakes abandon incubating females. The precocial chicks have good down coverings and some subcutaneous fat, which allows them to dive in icy waters almost from hatching. Northern seaducks are partially migratory, moving south ahead of the sea ice in winter.

Harlequin Duck *Histrionicus histrionicus*

Drake Harlequins are the most colourful of the non-Eider seaducks, with blue-grey head marked by white spots and crescents, and rufous streaks. The rest of the body is blue-grey, with white streaks and large rufous patches on the flanks. Females are much drabber, being mottled brown and white overall, with white patches around and behind the eye. Eclipse drakes are as the female but darker. Harlequin are river ducks, rarely being seen on lakes or flying over land: though Iceland's Lake Mývatn is a well-known spot for the ducks, even there they are usually seen at the river outlet where they feed on black fly (Simulium spp). The ducks nest on islands in rivers. Those living close to the sea often feed at sea, taking gastropods, crustaceans and amphipods.

Harlequin Ducks have a curious distribution, breeding in south-west and east central Greenland. East of Greenland the ducks breed in Iceland (the only place in Europe where the birds may be seen), but are then absent from most of Eurasia, appearing again in Chukotka, Kamchatka and around the northern shore of the Sea of Okhotsk, in southern Alaska, Yukon and North-West Territories. West of Greenland the ducks breed in southern Quebec and Labrador. In winter the ducks are resident or partially migratory, those that move being seen in the Aleutians, the north-eastern seaboard of the USA and on Hokkaido.

Long-tailed Duck *Clangula hyemalis*

Status: IUCN: Vulnerable (rapid decline in Baltic Sea winter numbers for unknown reasons). Drakes are unmistakable, with long, upcurved tail feathers. The head is white with a black patch on the cheek, the rest of the plumage black-and-white apart from a brown mantle and back. Females have brown-smudged white heads and brown upperparts. The breast is paler brown, the underparts white, though some birds have pale brown bellies. Female tail feathers are much shorter. Eclipse males are much whiter, particularly on the breast and upperparts. Females are much paler. Male Long-tailed Ducks moult again before breeding, being the only duck with different bonding and breeding plumages.

An earlier name for Long-tailed Duck was Oldsquaw. This was apparently a reference to the noisy drakes, whose call (a loud yodelling) reminded early travellers of the chatter of elderly Inuit women. In this age of political correctness, the name was considered inappropriate and changed, though it is still often used in North America. The scientific name for the bird translates as 'noisy winter bird'.

Long-tailed Ducks breed throughout the Arctic, including most of the Canadian Arctic islands. On Ellesmere Island the ducks vie with King Eiders for the title of most northerly breeding waterfowl. In winter the birds are seen in the Bering Sea, off eastern Canada, southern Greenland, around Iceland, north Norway and in the Baltic and North Seas.

a Male and female Harlequin Ducks brave the tumult of the Bering Sea. The pair were nesting on an island off Kamchatka's east coast.
b Female Harlequin Duck.
c Male Harlequin Duck.
d Incubating female Long-tailed Duck (and Red-necked Phalarope).
e Male Long-tailed Duck in winter plumage.
f Long-tailed Duck flock. It is autumn and the ducks are changing to winter/eclipse plumage.

Some ducks nest in tree holes, and will even accept nesting boxes. Given their webbed feet and general awkwardness, this is surprising. For species that nest this way (e.g. the Goldeneyes, Smew and Bufflehead of the northern ducks), their ability to colonise an area is dependent upon the presence of woodpeckers to excavate holes large enough for them to occupy.

In Iceland, where Barrow's Goldeneye use crevices in ancient lava flows as nest sites, the birds took to using nesting boxes on the sides of houses, despite it being extremely unlikely that any Icelandic bird had ever been so raised. When they leave the hole the chicks drop to the ground, occasionally falling 10m. There were once stories of the chicks riding away on their parents' backs or on their feet, but these are myths.

Bufflehead *Bucephala albeola*

The face, crown and neck of a breeding drake are black with a purple, green or bronze sheen. The rest of the bird is black-and-white. Females are much drabber, dark grey-brown above, with a prominent white patch behind the eye, and creamy brown below. There is a white speculum. Eclipse drakes are as females, but black rather than grey-brown and generally with a larger white eye patch.

Buffleheads are Arctic fringe species, breeding in central Alaska but more southerly across Canada, though occasionally seen north of the Arctic boundary as defined here. In winter the birds are seen to the south of the Aleutians, and off the Pacific seaboard.

Common Goldeneye *Bucephala clangula*

Drakes have a black head with a green sheen, and have a prominent white patch in front of the eye. The rest of the bird is black-and-white. Females have chocolate brown heads without a white patch. The rest of the plumage is grey-brown. There is a white 'speculum'. Eclipse drakes are as females, but some retain the white head patch. The second part of the species' name derives, as with the Long-tailed Duck, from 'noisy', drake Common Goldeneye having a modified windpipe which produces a loud, harsh bark.

Common Goldeneye breed in Fennoscandia and across Russia, but only to the north coast around the White Sea. Also breeds on Kamchatka. North American birds breed in southern Alaska, at the Mackenzie delta, then south towards Hudson Bay's southern shore and in southern Quebec and Labrador. In winter the birds move south of the breeding range, to both inland and coastal locations.

Barrow's Goldeneye *Bucephala islandica*

Barrow's Goldeneye is named for Sir John Barrow, the man behind the Royal Navy's expeditions in search of the North-West Passage in the 19th century. Drakes have black heads with a purple sheen, and a crescent-shaped white patch between eye and bill. Where the two Goldeneyes overlap, drakes can be most readily distinguished by the head gloss and the shape of the head white patch. Female Barrow's have a dark brown head and pale grey-brown body. Eclipse drakes are as females, but show more white on the upperparts.

Barrow's Goldeneyes breed in Iceland (mostly at Lake Mývatn where the population is declining) but are otherwise restricted to the Nearctic (the breeding birds of west Greenland are now almost certainly extinct). Nearctic birds breed in central and south Alaska, southern Yukon and British Columbia, and also in small numbers in northern Labrador. The reasons for this curious breeding distribution are unknown. Icelandic birds are resident, Nearctic birds move south in winter to inland and coastal sites.

a Male and female Buffleheads.
b, c Male Bufflehead, showing the iridescent colours of the head plumage of the breeding bird.
d Male and female Common Goldeneyes.
e Male and female Barrow's Goldeneyes.
f Barrow's Goldeneye flock.

Scoters

Scoters are black sea ducks with extraordinary bill shapes, the overall colour, bill and the fact that they swim with their tails low to the water are diagnostic. As with other ducks they are seasonally monogamous, males deserting the females once incubation begins, forming flocks with other males for the moult. In recent years the three Scoter species have expanded to five as very similar Nearctic and Palearctic forms have been separated into different species.

Common Scoter *Melanitta nigra*

Drakes are glossy black overall, though the underparts are paler. The bill has a prominent black protuberance at the base of the upper mandible, forward of which is a yellow patch, which extends beyond the nostril. The yellow coloration often extends on to the front of the protuberance and/or on its top, though the sides of the knob remain black. Females have a chocolate brown crown and nape, but the rest of the head is light brown. The remaining plumage is grey-brown. Males do not have an eclipse plumage, but first-winter drakes are browner with a less prominent bill knob. The birds hunt by diving (to about 6m) in search of molluscs, chiefly the Blue Mussel (*Mytilus edulis*), other bivalves, crustaceans and small fish.

Breeds in Iceland and across northern Eurasia to the Lena delta, but only to the north coast west of the Urals. Wintering birds move to the North and Baltic seas.

Black Scoter *Melanitta americana*

Status: IUCN: Near threatened (population decline for various reasons).

Formerly considered conspecific with the Common Scoter, the American Ornithologists' Union divided the two into separate species on the basis of the drake's bill pattern and shape, and also on differences in the mating calls (each produces a single note, that of the Black Scoter being longer). The plumage of male Black Scoters is essentially identical to that of male Commons, but the bill differs markedly. The base protuberance of breeding adults is entirely yellow-orange and is much less knob-like, being significantly flatter. The coloration of the remaining bill is similar in extent. The bill of Black Scoters (male and female) is slightly more hooked than that of Common Scoters. Female plumage is essentially identical to that of female Common Scoters.

Black Scoters breed in Asiatic Russia east of the Lena delta (but not to the north coast) and on Kamchatka. Also breeds in west and south Alaska, and in southern Quebec and Labrador. Both species are found in the lower Lena valley but there appears to be no overlap of ranges and no evidence of hybridisation. In winter the birds move to the Bering Sea and coasts of Japan.

Surf Scoter *Melanitta perspicillata*

Drakes are dull (rather than glossy) black apart from a white patch on the nape. The bill is a marvel. The upper mandible is swollen both up and out. The culmen is black-feathered to the nostril, the rest of the upper mandible being orange-red, white and black. Female is as Common and Black Scoters but with a pale patch at the base of the bill. First-winter drake is as breeding bird but duller, and lacks the white nape patch and glorious colours of the bill.

Surf Scoters breed patchily throughout Alaska (though are rare in the north), the Yukon and North-West Territories, in southern Quebec and Labrador. Wintering birds move to the Aleutians and east and west coasts of North America.

a Common Scoters.
b Female Common Scoter.
c Female Surf Scoter. While the white patch at the base of the bill of the female Surf Scoter distinguishes it from the female of Black and Common scoters, the latter are virtually indistinguishable, though the ranges of the two do not overlap and so they are not seen together.
d Male Black Scoter.
e, f Male Surf Scoter.

Velvet Scoter *Melanitta fusca*

Status: IUCN: Vulnerable (rapid population decline for unknown reasons).

The drake has a glossy black plumage which gives the species its name apart from a white half-moon eye patch and white secondaries. The bill is black at the base, becoming yellow and occasionally ends with a red tip, though this is sometimes not at all prominent. There is a small knob at the base of the upper mandible. Females are light brown overall, with a small white patch below and behind the eye. First-winter drakes are as females but duller, lacking the white half-moon eye patch and with a duller bill. Diet is similar to that of the Common Scoter, with the same enthusiasm for Blue Mussels. When taking off from water the birds run at first then lift their feet and fly very close to the water before rising into the air (Photo 87a, b and g).

Velvet Scoter breeds in northern Europe and eastwards across Russia to the southern Taimyr, but only to the north coast west of the Urals. In winter the birds move to the North and Baltic seas.

White-winged Scoter *Melanitta deglandii*

Formerly considered to be a subspecies of a single species including the Velvet Scoter, but now considered to be separate. The adult male differs in having a deep brown flank (Photo 87e). The bill pattern also differs, there being a much more pronounced knob at the bill base, extending over the nostril, the bill colour being red, and the feathering of the bill extending further both on the bill sides and on the culmen. However, males retain the half-moon white patch at the eye. Females are essentially identical to the female Velvet Scoter.

White-winged Scoters breed across North America from Alaska to the western shores of Hudson Bay, but only to the north coast near the Mackenzie delta.

The separation of the Nearctic and Palearctic birds into different species leaves the problem of where the birds of east Siberia sit. Covering the area from the Yenisey River to the Bering Sea, and extending into Kamchatka this population was originally considered a subspecies of the Velvet Scoter – *M. f. stejnegeri*. But in terms of bill pattern, these birds are much more similar to those across the Bering Sea and so may be a subspecies of *M. deglandi*. Further work is required to address this issue.

In winter birds from both North America and east Siberia move to waters off Japan, the Aleutian islands and the west coast of the United States.

a, b Male Velvet Scoters taking off from water.
c Velvet Scoter couple.
d Female Velvet Scoter with flounder.
e Male White-winged Scoter with bivalve.
f Female White-winged Scoter.
g Male Velvet Scoters.

Sawbills

The Mergus ducks are commonly known as sawbills because of their elongated, thin bills, which have serrations on the mandibles in order to grasp fish and hooked tips to aid their capture. There are three Arctic and subarctic breeders, though a fourth species, the Hooded Merganser *Lophodytes cucullatus*, while essentially a temperate bird, has recently been increasingly its range to the north.

Mergansers are the only ducks capable of catching fish, but they also take other prey, including amphibians, molluscs, crustaceans and even small mammals.

Red-breasted Merganser *Mergus serrator*
Drakes are handsome birds with dark green, glossy heads and necks and a prominent crest on the nape. The upperparts are black and white, the underparts grey and white. Females have a brown head and neck, also with a nape crest. The upperparts are grey, the underparts paler grey. The 'speculum' is white, split by a thin black bar. Eclipse drakes are as females. When diving, Red-breasted Mergansers use their wings as well as their feet. The birds will also feed cooperatively, an unusual trait, corralling a shoal of fish and driving it into shallow waters.

Red-breasted Mergansers breed on west and east Greenland, on Iceland and across Eurasia, though rarely to the north coast. Breeds on Kamchatka. Also breeds across North America, often to the north coast and including the southern part of Baffin Island. Most birds are resident or partially migratory, though western American birds move along the west coast of the United States.

Goosander *Mergus merganser*
Goosanders, occasionaly called Common Mergansers, are primarily freshwater rather than marine ducks. Drakes have a dark green head and neck with a green sheen. The upperparts and tail are dark greyish black, though the inner wings are white. The underparts are white. Females have brown heads and neck, light grey upperparts and grey-smudged white underparts. The 'speculum' is white. Eclipse drakes are as females but retain the completely white inner wing.

Goosanders breed in Iceland and across Eurasia, though only to the north coast in Fennoscandia; also breeds on Kamchatka. In North America the birds breed across the continent, but they are essentially subarctic. They are rare in west and central Alaska. They do, however, breed on the southern shore of Hudson Bay. Wintering birds are seen in northern France and the Low Countries, central Asia and eastern China, and off both coasts of the United States.

Smew *Mergus albellus*
Drakes are beautiful, with white heads apart from a large black eye patch and a black V on the nape. There is a small crest on the crown. The upperparts are black and white, though the tail is grey. The underparts are white and pale grey. Females have a chestnut crown and a similar black eye patch, but an otherwise white head. The rest of the plumage is grey, black and white above, white below. Eclipse drakes are as females, but paler and with a white crest.

Smews breed in southern Fennoscandia and across Russia to the Lena delta, but always essentially subarctic. Also breeds in southern Chukotka and northern Kamchatka. Wintering birds are resident or partially migratory.

a Red-breasted Merganser pair silhouetted by evening light among the ice of Hudson Bay.
b Red-breasted Merganser pair.
c Female Goosander.
d Male Goosander.
e Male Smew.
f Female Smew.
g Smews among frazil sea ice.

Raptors

Raptors are superbly equipped for the task of locating and catching live prey. They have talons for gripping fast-moving prey keen to make its escape, and hooked beaks for tearing flesh. The wings can be broad and long to allow the birds to soar with minimal effort as they seek relatively slow-moving animals or carrion, or tapered to allow very fast flight, the better to overhaul speedier prey. Raptors are solitary in the breeding season but sometimes become gregarious in winter, when congregations of birds and communal roosting of several of the larger species may occur, though most falcons do not share this behaviour, being solitary or in mated pairs at all times. The mating displays of some raptors often involve a form of 'sky-dance', in which the male circles the chosen nest site. There may even be mutual dancing, the pair touching or even linking talons. Sky-dancing is less common in falcons, though the 'high circling' flight of males above the nest site is an equivalent. Courtship-feeding is also common. In general the pair bond is monogamous and may be long-lasting, though in some raptors it appears to be seasonal. Raptor nests are often refurbished and re-used annually and can become huge, the sites highlighted by the streaks from years of defecation. By contrast, falcons do not make nests, either usurping the nests of other species or settling for a minimal scrape in the detritus of a cliff ledge. Gyrfalcons may actually lay eggs onto a bare cliff ledge.

The larger raptor species lay 1–3 eggs, the falcons 3-4. These are incubated by the female, which is fed by the male. The chicks have minimal down and are altricial or semi-altricial. They are fed by the female at first, with food brought by the male. The female also broods. Later, when the chicks have down and can thermoregulate, both birds will feed the young, though in general the female still does the bulk of the feeding, often with food brought by the male. The chicks are independent soon after fledging. Raptors take a long time to mature, and often do not breed until they are two or three years old, though the falcons may, occasionally, breed at one year. Steller's Sea Eagles do not breed until they are at least four or five years old, and perhaps even as late as eight years.

Apart from the harriers, hawks and some falcons, the sexes are similar in plumage. All species except the Osprey are sexually dimorphic and, interestingly, this is the reverse of the norm, with females being larger than males. This trait is shared by other avian predators such as owls, frigatebirds and skuas, though to nothing like the same extent in the latter two groups. It is not entirely clear what the basis for this reversed sexual dimorphism is: the size of prey taken by the two birds (smaller males are more agile and can catch smaller prey) is clearly important since it reduces interspecific competition, but why it should be the females that are larger (rather than the males as in almost all other avian groups) is not understood.

Breeding and winter plumages are similar in all Arctic raptor species.

Osprey *Pandion haliaetus*
Adults have long legs, large feet and long, narrow wings. The head is white with a thick, dark brown stripe from the shoulder through the eye. Ospreys are countershaded, white underparts camouflaging the bird against the (often cloudy) sky, and darker upperparts as camouflage if viewed from above against the water. Almost exclusively piscivorous – Ospreys are equipped with small spikes on their toes to aid gripping of prey.

Ospreys are subarctic, breeding as far north as the timberline across Eurasia. In the Nearctic they breed in southern Alaska (but are rare) and at the timberline across Canada. This distribution is the widest of any raptor. In winter Eurasian birds move to north Africa and south Asia, Nearctic birds to southern US states.

Eagles

True Eagles

The Aquiline or 'true' eagles (also occasionally called the 'booted' eagles because of their feathered tarsi) are a group of 30 or so species which include the genus Aquila, the largest and best known. Of these only one, the Golden Eagle, is a northern dweller, occurring at the Arctic fringe. Even though it can be difficult to assess the size of a solitary bird in the sky, it is unlikely the observer will misidentify the eagle, although there may occasionally be problems where the range of the species overlaps that of the few other raptors of the far north. These problems are noted below in the descriptions of the three species which may overlap.

Golden Eagle *Aquila chrysaetos*
Magnificent birds with long wings (the span may be 2.3m) that are narrower at the base than at the tip, and a long, fan-shaped tail. Adult have a pale rufous-brown head and nape (the golden colour which gives the species its name), the rest of the plumage being mottled brown and dark brown, the wings grey-brown. The sexes are similar. The birds of north-east Siberia and Kamchatka (a subspecies) are smaller and darker than European/western Russian birds, as are those of the Nearctic. Immature birds show a diminishing white wing-patch and white base to the tail (Photos 93a and e). While all white coloration may be gone at 3-4 years it often takes 6-8 years to acquire full adult plumage. Adult birds easily distinguished from adult Bald and White-tailed Eagles; from juvenile Bald Eagle by shape of tail, the frequent white mottling and lack of feathering on tarsi of the Bald Eagles; by tail shape and shape of rear wing (S-shaped on Golden, straighter on White-tailed) and frequent white mottling on White-taileds.

Feeds chiefly on mammals, usually smaller species such as marmots, ground squirrels and hares, but occasionally very large prey – there are stories of old and sick Reindeer being preyed upon, and of eagles taking Dall's Sheep and young deer – and on birds such as grouse, ducks and seabirds, but again including larger birds such as cranes. Will also feed on carrion. The birds hunt by patrolling their territory at height using their phenomenal eyesight to scan for prey, or by flying closer to the ground in the hope of flushing prey. Occasionally an eagle pair will hunting cooperatively with its mate, one eagle flying close to the ground, perhaps even walking, to flush prey which the second then takes. There are also instances of a pair using a similar tactic on other species which harass them, in one case an eagle allowing itself to be attacked by a Merlin which was nesting close by, distracting it so the second eagle could swoop on the unsuspecting falcon.

Golden Eagles use traditional nest sites on cliffs or, occasionally, in trees, and these may be refurbished annually, eventually becoming enormous: there are instances of nests 5m high and over 3m wide being claimed. 1-3 eggs, but usually 2, are laid. Incubation begins with the laying of the first egg meaning chicks hatch at different times. In general only the oldest chick will survive to fledging (on about 80% of occasions) as it will outcompete or even kill its siblings.

Golden Eagles breed in Fennoscandia, though are absent from the far north, then more southerly across Russia. In the Nearctic the eagles breed in Alaska (though are common only in the central state), to the Mackenzie delta, then more southerly, breeding around southern Hudson Bay and across southern Quebec and Labrador. Northern birds move south to join resident southern eagles in winter.

a, e Juvenile Golden Eagles.
b-d Adult Golden Eagles.

Fish Eagles

Perhaps not surprisingly given that the Arctic is an ocean (albeit a frozen ocean over much of its area) surrounded by land, that ocean linked to the Atlantic and the Pacific, 3 of the world's ten fish or fishing-eagles breed to the Arctic or Arctic fringe.

Bald Eagle *Haliaeetus leucocephalus*

A magnificent bird, the Nearctic equivalent of the White-tailed Eagle. Adults have a white head (the origin of the name) and white tail, but are otherwise dark brown, with a pattern of scalloping from pale feather tips (Photo 95c). Sexes are similar. Juvenile eagles do not acquire the adult colouring until they are 3-5 years old. In early young immatures there is a significant amount of white in the plumage which aids distinguishing the birds from adult Golden Eagles. Northern birds are larger than their cousins of the southern US states, the two being considered subspecies.

Primarily piscivorous but an opportunistic feeder, taking both terrestrial and aquatic mammals (e.g. hares and muskrats) as well as birds (primarily waterfowl and gulls, but is known to have taken Great Blue Herons (*Ardea herodius*)). Also feeds on carrion and is not an infrequent visitor to garbage dumps in Alaska. Hunts both by flying slowly over probable prey sites and from perches, and will pirate food from other Bald Eagles, and also from Osprey and herons, both of which are, in general, better at fishing but cannot defend themselves against the eagles. Also makes piratical attacks on Peregrines and even on Sea Otters and Coyotes.

Nests primarily in trees, and often re-uses nests which may actually be refurbished prior to the birds migrating as well as when they return to the breeding site. One nest known to have been used continuously over several decades grew to be almost 3m in diameter and over 3m high and weighed an estimated 2t: the tree then blew down in a storm. In areas where trees are absent, will nest on cliffs or even on the ground provided the site is elevated to give the incubating bird good visibility. 1-3, but usually 2, eggs are laid. Asynchronous hatching means first chick hatched usually outcompetes later hatchlings which may die of starvation.

The Bald Eagle is the emblem of the United States, which even sadder when numbers declined sharply as a result of subsidised shooting and the widespread use of organochlorines in the continental US. In Alaska the use of pesticides was minimal, but a bounty on eagles set in 1917 – the result of lobbying by fishermen and fox trappers – was not lifted until 1953. Although the bounty led to ill-advised slaughter, the state remained a stronghold of the species and continues to do so. Once hunting and DDT were banned the population recovered quickly: in Alaska the population increased by about 70% between the late 1960s and the late 1990s. Alaska and the Canadian province of British Columbia are the strongholds of the species with the population now probably close to the carrying capacity of the environment.

Bald Eagles breed in central and southern Alaska, including the Aleutians, and across North America, but rarely north of the timberline. In winter the birds move to the continental United States, though birds in southern Alaska are resident.

a, c Adult Bald Eagles.
b Juvenile Bald Eagle.
d Adult Bald Eagle at nest.
e This adult Bald Eagle was photographed at Alaska's Tracey Arm, a birthing site for Harbour Seals. The eagle was waiting for the opportunity to grab a new born or stillborn pup, or to feed on after-birth.

C

White-tailed Eagle *Haliaeetus albicilla*

Status: CITES: Appendix I.

Very large, with a wing span to 2.4m. White-tailed Eagles are Europe's largest raptor, equalling the Lammergeier (*Gypaetus barbatus*) in weight. The eagles have a long neck and a huge bill (Photo 97c). The largest birds are found in west Greenland (some experts believe these to be a subspecies, but there is no consensus on this), a decline in size then being observed moving eastward across the range. From personal observation the birds at the eastern extremity of the range are paler (Photos 97a and b) than those from the west (Photos 97c and d – northern Norway – and Photo 97e – Finland). Adults are uniformly mid-brown, with darker flight feathers and paler head and breast. The feathers of the upperparts have tips giving a scaly appearance (Photo 97d). The tail is white. Immature birds are darker, but with white mottling, this variable between birds and with age. The tail is brown, the adult white tail not being achieved until the bird is five years old.

Chiefly piscivorous but takes a remarkable range of prey, including carrion. Hunts from a perch, by soaring above likely prey sites or by flying low across water ready to snatch fish at the surface. Will also snatch birds from the water surface, species taken including divers and waterfowl on freshwater lakes or gulls and auks from the sea. On occasions will repeatedly attack a bird on the water's surface forcing it to dive each time until it becomes too exhausted to dive and is easily taken. Does not have the agility to take birds in flight, but will pirate birds carrying prey. Mammals taken range in size from rodents to young Reindeer. Despite the numerous legends and tall tales, there is actually only one documented record of an eagle having picked up a child. On 5 June 1932 a White-tailed Eagle snatched four-year old Svanhild Hansen, said to have been a particularly small child, at Leka in Norway and carried her for more than a kilometre to a ledge close to its nest site, almost 250m up a mountain. The girl was scratched, but otherwise unharmed. Also know to take reptiles and amphibians in southern parts of the species range. Carrion eaten includes marine species, the eagles having been seen feeding on carcasses of seals and cetaceans. The eagles are also known to feed on dead humans.

The eagles nest on cliff ledges or in trees, both adults contributing to the structure although in the main the male provides building material for the female to organise. 1-3, usually 2, eggs are laid, incubation starting after the first egg is laid so that hatching is asynchronous. However, siblicide appears much rarer than in Golden Eagles.

White-tailed Eagles breed in west Greenland, Iceland and across Eurasia, reaching the north coast in Fennoscandia and Chukotka. Also breeds in Kamchatka. Greenlandic and Icelandic birds are resident, west Eurasian birds overwinter around the North and Baltic seas, east Eurasian birds on the coasts of Japan, Korea and China.

d, e These shots were taken on sea ice in winter. Shot 'e' was taken at dawn, the rising sun illuminating the eagle's upper wing feathers.

Steller's Sea Eagle *Haliaeetus pelagicus*

Status: IUCN: Vulnerable (population decline due to habitat loss or degradation, and over-fishing).

The name of the bird in English commemorates the German doctor and naturalist Georg Wilhelm Steller (1709-1746) who in 1737-1742 participated in Vitus Bering's second Kamchatka expedition. Steller's name is associated with several bird and mammal species in recognition of his work on the expedition: on reaching Alaska during the trip, sensing history in the moment, Steller leapt ashore on Kayak Island and so became the first European to land on Alaska.

Steller's is the world's largest fish eagle and possibly the planet's largest bird of prey. Competitors for the title are the Harpy Eagle (*Harpia harpyja*) and the Philippine Eagle (*Pithecophaga jefferyi*), but with some females weighing up to 9kg, standing a metre tall and with a wing span of up to 2.9m the Steller's is at least comparable with both. The Steller's huge bill is without doubt the largest of any raptor and is reputed to be able to take off a man's finger with a single bite, though most experts discount this as a myth. Photo 99b illustrates the size of an adult Steller's by comparison to a White-tailed Eagle.

Adults have a white tail, legs and forewing (the latter explaining the Russian name for the species – White-shouldered Eagle), and occasionally a white forehead. The rest of the plumage is mottled dark brown and black. Immature birds do not acquire full adult plumage until they are six or seven years old.

The diet is almost entirely salmon when the fish is available, hunting being from a perch or during low, slow flying over a spawning river mouth or river. Should salmon not be available other fish may be taken as the species is almost exclusively piscivorous. However, waterfowl and gulls may be taken from the surface, and small mammals, including Arctic Foxes on land and seal pups at the water's edge may also be consumed. The eagles will also scavenge at the tide line, and regularly make pirate attacks on immature eagles and other adults.

The nests of Steller's Sea Eagles are re-used annually, new material being added. While cliff nests are not unknown, most nests are in trees and may sometimes be so vast – they can comfortably be the size of a standard car – that they overwhelm the tree, causing major branches (or even the entire tree) to break so that the nest falls. One or two eggs are laid, but asynchronous hatching ensures one chick of a two egg clutch is larger and this inevitably results in siblicide so that a single chick is raised.

Steller's Sea Eagles breed on Kamchatka, along the coasts of the Sea of Okhotsk and on northern Sakhalin Island. In areas where geothermal heat keeps lakes and rivers ice-free in winter – this chiefly on Kamchatka which forms part of the Pacific 'Ring of Fire' and so is volcanically activity along much of its length – birds are resident or move over short distances in response to local snow and ice conditions. Other birds move south to the area around Vladivostok and to the Japanese island of Hokkaido.

a, d Adult Steller's Sea Eagle.
b Adult Steller's Sea Eagle pursuing a White-tailed Eagle.
c Steller's Sea Eagle chick, probably about one week old. The chick's breakfast, just visible behind and left of it, was the same as the author's – salmon. The chick was not enthusiastic about having its picture taken, but was much more enthusiastic about discovering whether its bill really was strong enough to remove a man's finger.
e Immature Steller's Sea Eagle.

b

c

e

Hawks, Buzzards and Harriers

Taxonomists find it relatively straightforward to group the various species of raptor into genera and to group these into sub-families, a task made somewhat easier by the advent of DNA analysis of the various species. For the general reader the task is more complicated as despite the names hawk, buzzard and harrier apparently referring to three raptor types with very different characteristics in terms of wing shape and hunting techniques, the common names for species may sometimes add confusion, there being both buzzard-eagles and harrier-hawks, as well as the added problem of the species named the Rough-legged Buzzard in the Palearctic being called the Rough-legged Hawk in the Nearctic. For the Arctic traveller, apart from the minor difficulty of the common name of *Buteo lagopus*, life is easier, there being a single species of hawk, buzzard and harrier, each of which conforms to the characteristics that would be expected of the name.

Rough-legged Buzzard (Rough-legged Hawk) *Buteo lagopus*

A large bird with broad wings and a short fan-shaped tail. Adults appear in two colour morphs. Darker birds have dark brown upperparts with some white patches on primaries and covert leading edges. The underparts are off-white, but with brown breast and belly and brown patches on the wings and wing edges. In general the head is brown, but streaked white (Photos 101a, c and d). Pale morph birds are significantly paler overall in all areas (Photos 101b and f). In general in the Palearctic the further east across Siberia the traveller moves, the paler the birds are. Dark morph birds are more usual in North America, but paler birds are seen. The colour variations – there are intermediate forms and both extremely dark (and almost all over dark) birds are known, as are very pale birds – are not associated with subspecies. Indeed, the three subspecies which are often referred to vary only marginally and then in terms of size. Sexes are similar.

Rough-legged Buzzards will hunt from a perch if one is available, but otherwise hunt by quartering their territories. They may hover above potential prey before making a short stoop on to it. The prey is usually small rodents (Photo 101c shows a buzzard with such a rodent), but larger species – Arctic Ground Squirrels, Weasels – will also be taken, as will birds – small passerines and the chicks of grouse and geese. There are also reports of the birds feeding on the carcasses of dead Caribou and seals. Also pirates prey from other species including harriers, Ravens and conspecifics. Interestingly, an experiment has suggested that the birds may be able to detect the urine/faeces of rodents in UV light and so may, at least, be able to detect areas of high concentration of prey. Whether Eurasian Kestrels (*Falco tinnunculus*) are able to do this is currently a matter of considerable debate and so it will be interesting to see if the ability is confirmed in the two species.

Rough-legged Buzzards nest on cliff ledges (Photo 101e) or in trees and will re-use a nest or have several nests which are used in turn. Mating displays are more muted than in other species, but may include a 'sky dance' (Photo 101d). 1-6 eggs are laid, the exact number seemingly dependent on rodent density. Hatching is asynchronous, but there is little evidence of siblicide, though it may be that if food is scarce the younger, weaker chicks starve.

Breeds in Fennoscandia and across Russia, to the north coast except on the Taimyr Peninsula; absent from all Russia's Arctic islands. Also breeds in Kamchatka. Nearctic birds breed across the continent and on the southern Canadian Arctic islands. Arctic birds are migratory, moving to southern Eurasia and North America, where there are resident buzzards.

Hen Harrier *Circus cyaneus*

Typical harriers with broad wings and a long tail. Adult males have blue-grey heads and upperparts and paler underparts. The primaries are black. The central feathers of the tail are silver-grey with white tips, the remaining feathers are whiter. Adult females are very different, having brown heads and upperparts, the latter streaked buff and darker brown. The underparts are cream, streaked brown, the streaking heavier on the breast. The tail is grey-brown with highly conspicuous pale and dark bands when it is spread (Photo 103c). The most noticeable feature of the female at close range is the owl-like facial disc (Photo 103b).

Hen Harriers usually hunt by quartering their territories searching for small mammals, to the size of young hares, or birds which are then taken from the ground. The fact that the raptors take young grouse chicks has led to illegal killing of the species in northern Britain. The birds will also land on the ground and search for prey or, alternatively, having chosen a spot which seems favourable will wait until prey presents itself. Hen Harriers are also very agile fliers and will take birds from the air, even chasing down prey on occasions. Also known to take reptiles and amphibians opportunistically, to make piratical attacks on other species, for instance Short-eared Owls, and to feed on carrion.

Hen Harriers have the most extreme sky dance of any raptor, the male making a series of undulating rises and falls as though following the curves of a cork-screw. The female may join in, though sometimes this is associated with a food pass. The female may approach the male, turning in the air to present her talons. The male may lock onto these and the pair can then cart-wheeling across the sky in spectacular fashion.

Hen Harriers nest on the ground, it being an irony that the illegal killing of the birds in northern Britain is carried out chiefly on managed grouse moors where gamekeepers control the number of terrestrial predators, foxes and mustelids, which might otherwise prey on Hen Harrier nests. The nest is not elaborate, and into its shallow cup the female lays 4-6 eggs (though up to 8 in good prey years). Incubation is variable, sometimes starting with the first egg, sometimes later so that hatching is also variably asynchronous. Siblicide does not appear significant, but losses due to predation and also to trampling by ungulates reduce breeding success.

Hen Harriers are an Arctic fringe species, breeding in Fennoscandia, but not to the north coast, and then more southerly across Russia. Also breeds on Kamchatka. In winter the birds head for southern Europe and central Asia.

Northern Harrier *Circus hudsonius*

For many years it was considered that the Nearctic Northern Harrier was a subspecies, *C. c. hudsonius,* of the Hen Harrier, though some experts were of the view that they were separate species on the basis of differences in male plumage and wing length. In early 2016 it was decided to split the two when DNA analysis identified significant differences in support of the morphological differences.

Male Northern Harriers have darker upperparts than male Hen Harriers, and both the head and breast show pale rufous-brown spotting and streaking (Photo 103d). Females are as Hen Harrier females.

Foraging methods are the same as those of Hen harriers, but there are instances of Northern Harriers taking waders from the water surface and also of taking fish on occasions. Breeding behaviour is also very similar.

Northern Harriers are more Arctic than their Palearctic cousins, breeding in eastern Alaska, Yukon and western North-West Territories to the north coast, then more southerly to Hudson Bay and into Quebec and Labrador. In winter the birds move to the continental USA.

a, e Male Hen Harriers.
b, c Female Hen Harriers: female Northern Harriers are similar.
d Male Northern Harrier.

c

e

Northern Goshawk *Accipiter gentilis*

A boreal species and, therefore, one which occupies the Arctic fringe as it is rarely seen beyond the tree limit. Adult Eurasian males have grey heads with a distinct white supercilium. The upperparts are grey, occasionally washed brown (Photo 105a, the bird to the right, engaged in a territorial dispute with a juvenile). The underparts are white, thinly barred with dark grey. Females, which are significantly larger than males, are generally paler with the supercilium even more pronounced (Photos 105d and e). Juveniles are much browner, their underparts more spotted/speckled than barred (Photos 105a, bird to the left, and b). In far eastern Russia some Goshawks can be almost white and, if not perched so that the long legs are visible, may be confused with Gyrfalcons (until they fly). North American Goshawks are a subspecies (*A. g. atricapillus*) which are darker overall, but particularly on the head which makes the pale supercilium much more pronounced. Adult males may also not show the brown wash, though in the change from juvenile to adult male many birds go through a sub-adult phase when the brown wash is evident. Juvenile Nearctic Goshawks are very similar to their Old World cousins.

Goshawks are formidable combining astonishing flying agility, particularly evident when they thread themselves at speed through dense forests, with a lethal opportunism when it comes to defining prey. Eurasian birds take mammals and birds to the size of hares and Capercaillie, all taken by swift, aggressive flight which may be from a hidden perch in a surprise attack, or by a falcon-like stoop if the hawk is prospecting the area for a meal. In Britain Goshawks have been linked to the decline in numbers of both Kestrels, which may be killed as they hover and so are unaware of attacks from above, and Merlin. In North America Goshawks take squirrels and hares, and a variety of bird species including grouse. The hawks are also known to feed on carrion.

Nests are constructed in trees. A pair will usually have several which they will use in alternate years, each being re-used. Interestingly, while the female is responsible for the construction of new nests it appears the male usually refurbishes existing nests. This may be because the pair bond may not be maintained in successive years, so males will hope to win a female by offering a good territory and fine nesting opportunities. 1-6, but usually 3 or 4 eggs are laid. Though not synchronous, hatching is not as prolonged across the brood as in other eagles and hawks suggesting that incubation is either wholly or partially delayed. There is no evidence of significant siblicide, but, of course, if food supply is limited the weaker chicks may starve.

Goshawks breed to the treeline across both the Palearctic and the Nearctic. Most of the population is resident, but some birds move south in winter.

a Juvenile (left) and male (right) Northern Goshawk.
b Juvenile Northern Goshawk.
c Male Northern Goshawk.
d, e Female Northern Goshawks.

Falcons

Members of the Falconidae family are a successful group of hunters which have colonised every continent apart from Antarctica. The family is more extensive than a first thought might suggest, covering caracaras, forest-falcons and falconets, as well as the 'true' falcons which would spring to most people's mind on hearing the name. The true falcons include the Peregrine, arguably the most familiar species. Peregrines are also one of the most widespread of birds, sharing with Barn Owls (*Tyto alba*) the distinction of nesting on six continents. True falcons are identified by their 'swept-back' wing, a narrow, pointed wing designed for high-speed flight. True falcons do not build nests, either using an old nest constructed by another species, or laying their eggs onto a cliff ledge with little or no substrate. Most of the true falcons are temperate zone dwellers, but three are Arctic dwellers.

Gyrfalcon   *Falco rusticolus*
Status: CITES: Appendix I.

One of the iconic species of the Arctic. A magnificent bird, the world's largest falcon, and with a body length of up to 60cm and wing span to 1.6m as large, or marginally larger, than some accipiters. Gyrfalcons come in three distinct colour morphs (though with some intermediate forms as well). The darkest birds have dark slate-grey heads and upper parts, with underparts which may be almost as dark, or paler, with paler patches on the chin. The leg feathering may also be paler. In some dark birds the head is so dark that there is no hint of a moustachial stripe, though some birds (as in Photo 107d) have paler cheeks allowing the moustache to be visible. Grey morphs have paler upperparts with white speckling and even some darker speckles. The underparts of grey Gyrs are white with black/dark grey speckling. Some grey morphs have slate-grey crowns, the rest of the head a similar colour, but with white cheeks and throat, the cheeks allowing a visible moustache. There may also be a white supercilium. The grey adult in Photo 107b shows such a head colour. Other grey birds may have paler grey-white heads with with some dark streaking (from feather shafts) and a less distinct moustache (Photo 107c).

The most beautiful Gyrs are the white morphs of the High Arctic. Some of these birds are almost pure white, with minimal black barring on the upperparts (Photo 107f) and paler barring on the underparts – small wonder that many travellers consider them the ultimate tick on their bird list. An intermediate form can be seen in Iceland where many of the Gyrs are extremely pale, but never truly white (Photo 107e). In winter Gyrs from north-east Greenland may migrate to Iceland and since many of these birds will be white morphs, interbreeding with the essentially grey morph local gyrs may explain the intermediate form.

Gyrfalcons hunt birds, primarily Lagopus species which share their far north landscape. Coastal Gyrs will also hunt gulls, while inland Gyrs will hunt ducks and geese. Smaller birds will also be taken if available. Mammals are also taken, particularly Arctic Ground Squirrels and hares. Hunting technique varies with terrain. Gyrs will quarter the ground in the manner of harriers, while ride the air currents of ridges and then stoop on prey (see the following pages on Peregrines regarding the speed of stooping falcons). They will also hunt from perches using surprise as a tactic. Breeding Gyrs utilise the old stick nests of Ravens and Rough-legged Buzzards, but often lay eggs on bare ledges. Laying is very early, often when snow is still lying so that the female must sit on the clutch (1-4, usually 3 or 4 eggs) more or less permanently to prevent chilling. Hatching is usually asynchronous, but there are few reports of aggressive behaviour between siblings.

Gyrfalcons breed in Greenland (both west and east coasts), Iceland, across Eurasia, but not on Russia's Arctic islands; they also breed on Kamchatka. Breeds across North America, including Canada's Arctic islands as far north as Ellesmere Island. Gyrs are resident, but move locally in search of prey.

a-c Grey morph Gyrfalcons, Norway.
d Dark morph Gyrfalcon, Norway.
e Intermediate morph Gyrfalcon, Iceland.
f White morph Gyrfalcon and nestlings, Canadian High Arctic.
g Gyrfalcon family, north-east Siberia. The adult is white morph, but the chicks are grey.

Peregrine *Falco peregrinus*

Status: CITES: Appendix I.

One of the world's best-known birds, though this has less to do with its beauty and speed than the rapid decline the species underwent in the 1950s and 1960s as a result of the widespread use of organochlorine pesticides (specifically DDT), and the reintroduction and conservation programmes that followed, which raised both numbers and public awareness. Peregrine numbers are now much healthier in western Europe and North America, though the situation is not as good in all parts of the bird's range.

Peregrines are among the most charismatic of birds. Adults have a dark blue-grey crown and nape, and a cheek patch of the same colour, a feature usually called a moustachial stripe (often shortened simply to moustache) though it does not actually extend below the bill. The upperparts are blue-grey, the underparts pale cream or buff, barred with dark brownish black. Females, which are significantly larger than males are generally more heavily marked. There are numerous subspecies of Peregrine, the most interesting from a point of view of the Arctic traveller being the 'Tundra' Peregrines of North America and Asian Russia. *F. p. tundrius*, which occupies northern Alaska, Canada's northern mainland and southern Arctic islands, and Greenland is larger and paler than nominate birds (which are seen in western Europe, including Fennoscandia), the most pronounced pale coloration being on the breast (Photo 109d). The Asian Arctic Peregrine, *F. p. calidus*, is similar to *tundrius*, but tends to have a darker crown (Photo 109b).

While the fast stoop for prey is the most spectacular hunting technique, it is not the only one employed by Peregrines as they will make surprise attacks from perches and also quarter the ground in the manner of harriers in the hope of flushing prey into the open. Stoops are from various angles, lower angles looking as though the falcon is chasing down its prey. At higher angles the speed of the Peregrine is legendary and has led to the falcon being classified as the fastest creature on the planet with speeds in excess of 300km/h being claimed for Peregrines accompanying human sky divers in near vertical stoops. While these tests certainly imply the birds are capable of such speeds, several things are worth considering. Firstly the time required for the falcon to acquire these high speeds, and the distance travelled to do so preclude any possibility of them being reached in true hunting: the prey would be long gone before the Peregrine arrived. Peregrines that do not 'bind on', i.e. grab and hold, to their prey rake it with their talons to inflict serious, perhaps fatal, damage and then retrieve the prey in the air or from the ground. At such speeds the Peregrine would likely incur life-threatening damage when striking. In practice much lower speeds have been recorded for attacking Peregrines, speeds of about 180km/h and usually much lower. And finally, as the Gyrfalcon is heavier and so accelerates faster in a steep dive it is not surprising that hunting Gyrs have been recorded at higher speeds (around 200km/h).

Peregrines take available prey and often specialise, so that coastal falcons will take, for instance, mainly auks, while inland birds will favour ducks or doves. In all cases, of course, the falcon will opportunistically take whatever prey presents itself.

Peregrines nest at traditional sites, in the Arctic these inevitably being cliff ledges. 1-6 eggs , usually 3 or 4, eggs are laid, and there is little evidence of competition between siblings resulting in siblicide, though younger or weaker chicks may starve.

Peregrine Falcons breed on Greenland and across Eurasia, including the southern island of Novaya Zemlya, and also on Kamchatka. In North America they breed throughout Alaska and Arctic Canada, including the southern Arctic islands.

a, e, f Nominate adult Peregrine Falcons in Scandinavia.
b Breeding F. p. calidus with egg clutch in northern Siberia.
c Breeding F. p. tundrius with egg clutch in northern Canada.
d Adult F. p. tundrius, northern Canada.

d

f

Merlin *Falco columbarius*

The Merlin is a curious falcon, among the smallest of the true falcons, but one which sits outside the standard 'four group' classification of them. It is also a cold climate specialist when most of its true falcon cousins are birds of temperate zones. Only the very much larger Gyrfalcon compares as a cold specialist, though some Peregrine subspecies are, as we have seen, also Arctic dwellers.

Adult male Merlins have a blue-grey crown and auriculars, and well-defined white supercilium. A weakly defined moustache is usually visible. The upperparts are blue-grey covered with fine dark streaks. The throat is white, the remaining underparts being pale rufous-brown with darker brown streaks. The undertail is dark blue-grey with several (usually four) paler bands. Adult females have a brown head and upperparts, and a more pronounced moustache. The underparts are cream, heavily streaked dark brown. The undertail is dark brown with several pale bands. This description is for the nominate Merlin, the 'Taiga Merlin', the most northerly of three subspecies in the Nearctic. In the Palearctic there are six further subspecies, one breeding in Iceland (*F. c. subaesalon*), one in western Europe, including Fennoscandia and Russia to about 120°E (*F. c. aesalon*), another breeding across remaining Siberia (the East Siberian Merlin – *F. c. insignis*), and, finally, one on the Bering Sea coast of Chukotka – the Pacific Merlin, *F. c. pacificus*. In all cases the differences between these subspecies, in size and coloration, are marginal.

As with all true falcons, Merlins are built for fast flight, but they are also extremely agile and may chase down prey. However, most hunts are by surprise attacks, often from a perch chosen to give the falcon a wide view across its territory, though sometimes by quartering the ground and flushing prey. The prey is chiefly avian – in Iceland there were no rodents until man settled the island, and even now introduced mice form only a fraction of Merlin prey – but in Scandinavia where rodent numbers may rise spectacularly in so-called lemming years the fraction of mammals in the diet of Merlins rises substantially.

As with other true falcons, Merlins do not build nests. In North America utilise corvid stick nests in trees where these are available, but will also nest on cliff ledges, the female making a scrape in available substrate. In Iceland, where there are few trees and only one corvid, the Raven, old Raven stick nests are used if available. There are also few cliffs so the Merlins use flat sections on steep banks as nest sites. In some places there are deep trenches created by volcanic activity and Merlins use ledges in these if they are available. By personal observation, in one such trench Merlins were nesting only 200m or so from Gyrfalcons who were utilising another (rather more commodious) ledge: there was no evidence of aggressive behaviour between the falcons, and there is no evidence from Icelandic studies of Gyrs predating the smaller falcons. In Britain ground nesting of Merlins is common, and there are instances of this occurring in Scandinavia as well, though the use of corvid tree nests is more prevalent. One interesting aspect of Merlin biology has been the recent use of artificial nests and also the migration of Merlins into urban areas (the 'Prairie' Merlin (*F. c. richardsonii*) in southern Canadian towns). 1-7 eggs, usually 3-5, are laid, and as with the other falcons there is little evidence of siblicide among the chicks, though as might be expected with an Arctic breeding species, broods are prone to succumb to extended periods of bad weather.

Merlins breed in Iceland, Fennoscandia, and then more southerly across Russia, though are absent from Kamchatka. They breed in central and southern Alaska (but are rare throughout the State) and more southerly across Canada to Hudson Bay and in southern Quebec and Labrador. All northern birds are migratory, flying south to join resident birds.

*a Adult female Eurasian Merlin (*F. c. aesalon*).*
*b Adult male Eurasian Merlin with a Meadow Pipit (*Anthus pratensis*).*
*c Adult male Taiga Merlin (*F. c. columbarius*).*
*d Adult female Pacific Merlin (*F. c. pacificus*) with Yellow Wagtail (*Motacilla flava*).*
*e Male Icelandic Merlin (*F. c. subaesalon*)*
f Female Icelandic Merlin.

Grouse

Grouse are a group of 19 northern gamebirds. Several of this highly successful group have wide latitudinal ranges: (Rock) Ptarmigan are found from about 38°N to 82°N. Grouse are stocky birds, their size a reflection of their diet, particularly the winter diet. Boreal species eat the needles of pine and spruce. Though abundant, the needles have a very low nutritional value so huge quantities need to be ingested to provide the bird's energy requirements. Large crops and long intestinal tracts are therefore necessary.

In winter, tundra grouse moult into a dense white plumage offering high insulating qualities and excellent camouflage (both avian and mammalian predators share their winter quarters). The birds also grow horny appendages (pectinates) along the sides of the toes. Shaped like scoops, the pectinates break off in the spring. All but three of the grouse species grow these curious appendages: no other species of bird grows them. The exact purpose of the pectinates is still debated, with some experts believing that they aid perching. However, pectinates appear on non-forest dwellers, and their form and rigidity does not obviously convey any advantage to a perching bird. Much more likely is that the pectinates are an aid to digging in snow. Although some grouse migrate in winter, occasionally travelling considerable distances, in general all fly infrequently and are essentially terrestrial species, being resident or moving short distances to find better foraging. For most grouse, and certainly for the most northerly ones, winter brings extensive snow cover. Each day the grouse dig a burrow into the snow: on days of heavy snowfall they may dig several burrows each day or have to excavate the roof of their burrow to avoid becoming entombed. Pectinates would appear to be a solution to this constant need for digging: they are a seasonal snow shovel.

Snow burrows are excavated by both feet and bill, the bird digging a tunnel 0.6-4m long at the end of which it creates a chamber large enough to accommodate itself in comfort. Within the chamber there are three sources of heat. One is the bird itself, though the superb insulation of the winter plumage minimises heat loss. The second is exhaled air. The thick feathering of the nostrils ensures that moisture is condensed and returned to the bird, but heat nevertheless escapes. Finally, the bird excretes every few minutes throughout its stay in the burrow as the winter forage is extremely poor and waste is a high fraction of intake. These sources of heat mean that even if the ambient temperature falls to -40°C the temperature in the burrow does not go below -3°C and the bird remains snug. →

Ptarmigan (Rock Ptarmigan) *Lagopus mutus*

Breeding males have a dark brown or grey-brown head, mantle and back, each streaked darker. The breast is similar, blending to white underparts. The wings are white. There is a red comb above the eye. The tail has black outer feathers. Females are cryptic, mottled dull brown. In winter the birds are white, but retain the black outer tail feathers.

In summer the colour of the male is diagnostic as Willow Grouse males are red-brown, though females are extremely difficult to differentiate. In general Ptarmigan females are darker, but as there is an overlap of base colour this is not always a good indicator. The bill of Ptarmigan is smaller and thinner than the stouter bill of the Willow Grouse. Once in winter plumage bill size may still be used to distinguish the two species, but the easier characteristic is the black lores of the Ptarmigan.

Ptarmigan breed on Greenland, Iceland and Svalbard (the Ptarmigan of the island are a subspecies, *L. m. hyperboreus*, which is smaller and rare), and across Eurasia to Kamchatka, to the north coast. May also breed on Russian Arctic islands. Also breeds across North America, including all Canadian Arctic islands to northern Ellesmere Island. Resident.

a *Ptarmigan pair in winter plumage, female left and male right, northern Norway.*

b *Male Ptarmigan moulting to breeding plumage, Iceland.*

c *Female Ptarmigan in breeding plumage, Iceland.*

d *Male Ptarmigan of the subspecies* L. m. captus, *which inhabits northern Greenland and is probably the world's most northerly breeding bird. It was autumn when this photograph was taken, and the bird was moulting to winter plumage.*

e *Female Svalbard Ptarmigan in winter plumage.*

Willow Grouse (Willow Ptarmigan) *Lagopus lagopus*

Breeding males have red-brown head and neck, the mantle and back mottled dark. The breast is red-brown, streaked darker and merging into the white belly and vent. The outer wings are white, the inner red-brown. There is a distinctive red comb above the eye. Females are cryptic (Photo 115b), mottled dull brown. Some retain a white belly, and some have a red eyebrow. Wings and tail as male. Winter birds are white, but retain the black outer tail feathers, probably for signalling as the birds are gregarious in winter. Both Ptarmigan and Willow Grouse are notable for having feathered feet (Photo 115e), a characteristic shared only by the Snowy Owl. But the terrestrial nature of both species means that the feet are also feathered, the feathering encircling the toes to aid traction on snow.

Willow Grouse breed across Eurasia to Kamchatka, north to the coast and on the New Siberian Islands. Absent from Greenland, Iceland and Svalbard. Breed across North America including the southern Arctic islands. Resident in all areas.

→The tunnel and chamber lie close to the surface – the bird may push its head up through the snow while digging to measure depth, and always pokes its head through the chamber roof to check for danger before leaving the burrow. Once the bird has ensured their is no local danger it explodes from the burrow and immediately start feeding. Since it can accumulate enough material to sustain it for a day in 30-40 minutes of non-stop feeding the bird may spend 23 hours in the burrow in bad weather, emerging only when hunger forces it to.

In spring males moult to a distinctive breeding plumage, while females acquire a cryptic plumage – a necessity for these ground-nesting birds, which provide a high fraction of the vertebrate biomass of the tundra and boreal fringe and are important prey for many species. One such predator, the Gyrfalcon, is, in some places, almost entirely dependent on grouse, taking virtually no other prey throughout the year.

The conspicuous plumage of male grouse would appear risky in situations where some predators rely on grouse meat for survival. But the males must advertise in order to attract females and to show ownership of a territory to potential rivals. Advertising requires males to choose highly visible perches, but the male is playing a dangerous game, pitting its ability to outfly a predator against the chances of being chased down. If the male sees a mammalian predator in time, its rapid acceleration will take it clear. In areas with good patches of cover it can usually make it to cover before being overtaken by a pursuing bird; indeed, in areas where the shrub cover is abundant Gyrfalcons often do not breed. It is, however, a different matter on open tundra.

The males of the Arctic species tend to perform solo displays involving tail fanning, dropped wings and exaggerated strutting; the birds perform to intimidate rival males, since the holding of a territory is key for mating success. The males of some southern species take a more communal approach; 'leks' are created, displays grounds in which the males strut their stuff. Females visit the leks, observing the displays before making a final choice. Once mated a female grouse will lay a clutch of up to a dozen eggs, though 5-8 is a more usual number. The female alone incubates. Grouse chicks are downy and precocial, well capable of feeding themselves. Unlike their parents, the chicks feed chiefly on invertebrates, changing to a more vegetarian diet as they grow. They are fully grown by their first winter, but do not always breed the following spring.

Grouse are resident in many Arctic areas, though they will move if local food sources become too poor or unavailable as a result of bad weather (e.g. rain followed by frost, which seals food beneath an impenetrable coat of ice). This limitation on movement and the wide geographical range of the birds has led to the evolution of many subspecies; as many as 30 have been recognised for the (Rock) Ptarmigan by some experts, but usually the differences between these are marginal.

a Willow Grouse in winter. Note the absence of black lores, a distinguishing characteristic of the species.

b, c Female (left) and male (right) Willow Grouse in breeding plumage.

d, e Female (left) and male (right and 'f') Willow Grouse. The female is in winter plumage, but the males have not yet fully acquired theirs.

Spruce Grouse *Falcipennis canadensis*

A boreal species that can be seen to the timberline. Breeding males have a grey-brown head and upperparts, black neck and white-spotted brown underparts. The tail is dark brown with a distinct orange-brown terminal band. The tail is fanned during displays. There is a red comb above the eye. Females are cryptic brown. Winter plumage is as breeding.

Spruce Grouse breed in central and southern Alaska (but are rare in the west), and to the timberline across Canada. Resident.

Ruffed Grouse *Bonasa umbellus*

A boreal species that prefers deciduous woodland and so is not a true Arctic species, but may be seen by travellers enjoying Alaska's Denali National Park. Adults have two colour phases, with either rufous upperparts and white underparts with rufous streaking or black, or grey-brown upperparts and white underparts with grey-brown or black streaking. There is a crest on the head. The male tail, fanned in display, has a red-brown inner, grey centre and black and pale brown terminal bands. Winter plumage is as breeding.

Ruffed Grouse breed in central Alaska (but not elsewhere in the state) and across North America, though rarely to the timberline. Resident.

Hazel Grouse *Bonasa bonasia*

A boreal species that may be seen to the timberline. Adult males have grey upperparts and brown wings, each scalloped darker. The underparts are white, heavily marked brown. The throat is black with a white border. The tail is grey with thick black and thin white terminal bands. Females are similar but lack the black throat. Winter plumage is as breeding.

Hazel Grouse breed in southern Fennoscandia and to the timberline across Russia. Absent from Chukotka and Kamchatka. Resident.

Western Capercaillie *Tetrao urogallus*

A boreal species that can be seen to the timberline. Breeding males are magnificent birds with shiny blue-black plumage and white spotting on the belly. The wings are brown above, grey below. The tail is blue-black and fanned into a semi-circle at leks. There is a red comb above the eye. The bill is pale yellow. Females are cryptic brown. Winter plumage is as breeding.

Capercaillies breed in southern Fennoscandia, then more southerly across western Russia. Resident.

Black-billed Capercaillie (Spotted Capercaillie) *Tetrao parvirostris*

Similar to (but smaller than) the Western Capercaillie, with which it overlaps sympatrically in central Siberia. Adult males are as Western Capercaillie, but with white spotting on the upperparts and a black bill. Females are cryptic brown and similar to Western females. Winter plumage is as breeding.

Black-billed Capercaillie breed east of 100°E, but only to the timberline; Kamchatka is a stronghold. Resident.

a *Male Spruce Grouse.*
b *Female Spruce Grouse.*
c *Adult rufous Ruffed Grouse.*
d *Female Hazel Grouse.*
e *Male Hazel Grouse.*
f *Female Western Capercaillie.*
g *Male Western Capercaillie.*
h *Male Black-billed Capercaillie.*

Cranes

Cranes are the tallest of the flying birds, a creature seemingly too delicate to withstand the rigours of the Arctic. Yet two species are found in the north, and each is a true Arctic species. Cranes are omnivorous, opportunistic feeders, taking seeds, berries, invertebrates, amphibians and reptiles, and even small birds. Wintering Sandhill Cranes feed in fields, taking waste grain and potatoes, but the Siberian Crane is less adaptable, feeding only on wetlands. This specialism represents the greatest threat to the species, as wetland drainage along its migration routes and at its wintering sites have reduced feeding possibilities. The Chinese Three Gorges hydro-electric scheme and the proposed dam at the Poyang Lake outlet represent a considerable threat to the birds' prime wintering areas. Such severe hydrological threats are inevitably to the detriment of the cranes and, it is feared, may result in a continuing decline in species numbers despite attempts in Russia to stabilise, and then raise, the population. Siberian Cranes are not alone in being threatened, as the majority of crane species are on the IUCN List and all are either CITES Appendix I or II. In the case of the Siberian Crane the situation is not helped by females occasionally laying only a single egg rather than the 'standard' two, and the chicks vulnerability during the long fledging period.

In general cranes are monogamous. Though gregarious on migration and at wintering sites, they tend to be solitary while breeding (though the Sandhill Crane is an exception). Cranes are renowned for their 'dancing' displays, which all species perform (though the Red-crowned Cranes *Grus japonensis* of Japan are justifiably the most famous). The dances consist of wing-stretches, head-tosses and vertical leaps, and calling in unison with extended necks. The calling is also used for territorial claims. Most cranes breed at 3–5 years, but they are long lived. They make large nests in which, usually, 2 (perhaps 3) eggs are laid. Both birds incubate. Crane chicks are precocial, following their parents into feeding areas at an early age, but they are initially fed by both parents. The chicks do not develop flight feathers until they are 3–4 months old.

Sandhill Crane *Grus canadensis*
Status: CITES: Subspecies *nesiotes* and *pulla* are Appendix I.

Unmistakable. Adult have a red crown, forehead and lores. The rest of the head is white, the neck grey. The upperparts are a grey-flecked reddish-buff, the underparts pale grey. Sexes similar. Winter birds lose the reddish-buff wash on the upperparts.

Sandhill Cranes breed in coastal Chukotka east of the Kolyma delta and south to the Gulf of Anadyr, and across North America from Alaska to western James Bay. They are uncommon in northern Alaska, yet breed to the north coast along much of the Canadian mainland, and on Victoria and Baffin islands. Migratory, with Russian birds joining those from North America in the southern United States.

Siberian (White) Crane *Leucogeranus leucogeranus*
Status: IUCN: Critically Endangered (probable population decline due to threats of dam building in China on wintering population).
CITES: Appendix I.

Unmistakable. Adult are pure white apart from a red crown, forehead and lores, and black primaries, primary coverts and alula. The black feathering is often visible on the standing bird. Male, female and winter plumages similar.

Siberian Cranes breed in two areas, with a small population near the Ob River and larger one in northern Yakutia. The Ob birds migrate to the Iranian Caspian Sea and northern India, with hunting along the route (especially in Pakistan and Afghanistan) adding to the threats the species faces. The Yakutian birds move to China, to the area compromised by the Three Gorges scheme.

a, b Sandhill Cranes. The adult in 'b' is in breeding plumage and is dancing.
c–f Siberian Cranes. The chick in 'c' has just hatched in a two-egg clutch.

Shorebirds

Shorebirds (or waders) represent the largest group of Arctic birds, working the intertidal region of the Arctic coast, and the lakes and marshes created by the summer thaw, for insects, worms, crustaceans and molluscs. This animal material is taken together with some plant material, and occasionally other foods such as small fish. There are several shorebird groups (especially the calidrids) in which plumages are very similar between the species so the risk of confusion between them is high, particularly as the birds also generally have distinct breeding, wintering and immature plumages. In the species accounts that follow some pointers for identification are included, but it must be stressed that these are general: the identification of waders requires patience and perseverance.

Waders generally nest on the ground (though Green, Solitary and Wood Sandpipers use the old nests of other species, often high above the ground; Green Sandpipers do this frequently, sometimes at heights to 13m). The nest is little more than a scrape in the ground or a depression in vegetation, often with minimal lining. Between 3 and 5 eggs are laid. The eggs are large – as they must be to accommodate a chick which is born downy and precocial – and as a consequence represent a large energetic investment for the female; for some of the smaller calidrids (e.g. the stints) the clutch can weigh up to 90% of the female's body weight. Care of offspring differs between the species. In most it is shared by the parents. In some it is carried out by the female alone, while there are species in which the male takes responsibility. In species in which both birds incubate the eggs and care for the young, the female often abandons the chicks before they fledge, with the male continuing with care. Young shorebirds leave the nest within a few days of hatching and are, in general, self-feeding. Shorebirds breed in their second or third year.

Oystercatchers

Oystercatchers are noisy, gregarious birds (though more solitary at breeding sites). They are also unmistakable with their pied plumage (though the North American Black Oystercatcher differs in being all black). They use their long, brightly coloured, bills to probe beneath the surface of beaches to find molluscs attached to rocks. Bivalves are opened by inserting the bill between the two shells and severing the adductor muscle, which holds the shell closed. With limpets, and occasionally with bivalves, the bird may also hammer away with its bill on the shell, excavating a hole through which the flesh is extracted. Two oystercatcher species breed at the Arctic fringe and so may be seen by travellers to the region.

Eurasian Oystercatcher *Haematopus ostralegus*
Sexes are similar. Adults have a black head, black-and-white upperparts and white underparts. The bill is long and red or red-orange. The legs are also red-orange. In winter the birds acquire a white half-collar.

Eurasian Oystercatchers have a discontinuous distribution, breeding on Iceland, on the northern coast of Fennoscandia and around Russia's White Sea coast, and on Kamchatka. Some Icelandic and Norwegian birds are resident, but most northern birds head south, some as far as the Bay of Biscay, but most to coastal Low Countries and northern France. Kamchatka birds move to south-east Asia.

Black Oystercatcher *Haematopus bachmani*
Sexes are similar. Adults are black with a bright red bill and dull pink legs. Winter plumage is similar. Breeds on the Aleutian Islands and southern and south-east Alaskan coasts.

Mainly resident, but some birds move south along the western coast of the United States.

a, c Black Oystercatchers.
b, d Eurasian Oystercatchers.

Plovers

Plovers are stocky, short-billed birds with large eyes, and are characterised by their feeding method of 'look–run–peck'. The pecking is on land forays, or of crustaceans and molluscs on the shore. Some plovers also use a foot-paddling technique when feeding; the bird stands on one leg and paddles the water surface with the other, the paddling causing prey to move and betray itself.

In the tundra plovers (as the Golden and Grey plovers are known) the pair bond is monogamous and probably lifelong; initial pairing takes place after a display that involves the male running at the female (the 'torpedo run'), which looks far from subtle but seems to work. The ringed plovers have similar displays, though the pair bond, while monogamous, is seasonal. Things are different in the Dotterel. Females take the lead in mating, luring males with raised wing and fanned-tail displays. Female plovers usually abandon their eggs soon after laying to seek out another male for a second clutch, leaving the first male to care for eggs and young. Female Dotterel rarely share brood responsibilities.

Plovers are famous for displays aimed at drawing potential predators away from their nests and eggs. These displays usually involve the bird drooping a wing and calling plaintively as though badly injured as it heads away from the nest (Photo 123g). Human invaders seem especially gullible to the performance.

(Eurasian) Dotterel *Charadrius morinellus*
Breeding adults have black crowns above a white face. The upperparts are grey and grey-brown. The upper breast is also grey, separated from the rufous belly by a white band (Photos 123a and b). Females are brighter (and larger) than males. In winter the birds lose the rufous underparts and are duller overall.

Dotterels breed in Fennoscandia and across northern Russia to Chukotka, including Novaya Zemlya's southern island. Absent from Greenland, Iceland and Kamchatka. Has bred on the tundras of western Alaska. In winter the birds fly to the Middle East and North Africa as far west as Morocco. East Siberian birds may travel more than 10,000km during migration.

Lesser Sand Plover (Mongolian Plover) *Charadrius mongolus*
Adult has a 'highwayman' mask similar to Ringed Plover. The crown is grey, ringed with pale rufous. The throat is white, sharply delineated from a rufous breast that merges into white underparts. The upperparts are grey-brown, the tail is grey, dark grey and white (Photos 123c and d). In winter the back facial band and rufous colour are lost.

Breeds in Chukotka, Kamchatka and on the Commander Islands. Has also bred in western Alaska but is not an established species there. In winter the birds move to the Philippines, Indonesia and Australia.

(Common) Ringed Plover *Charadrius hiaticula*
Adult has a black 'highwayman' mask with bands above and below a white forehead. The crown and nape are grey-brown. There is a distinct white crescent by the ear, with a broad white collar, below which is a broad black chest band. The rest of the underparts are white (Photos 123e and f). Females are less boldly marked than males. In winter the black facial and chest bands are more subdued.

Breeds in Greenland, Iceland, Svalbard, Scandinavia and along the northern coast of Eurasia including the southern island of Novaya Zemlya and the New Siberian Islands; has also bred on St Lawrence Island, northern Ellesmere and western Baffin Island, where it is now thought to be established. In winter European birds move to southern Europe and North Africa, Asian birds to the shores of the Caspian Sea and the Middle East.

Semipalmated Plover *Charadrius semipalmatus*
Virtually identical to the Ringed Plover in summer and winter, but lacks the prominent white ear patch, having an indistinct crescent (and sometimes no white at all) (Photo 123h). The name derives from the webs between the inner and middle, and middle and outer toes.

Breed across North America and on Banks, Victoria and Southampton Islands, and southern Baffin Island. Wintering birds move to the southern Pacific and Atlantic coasts of the United States, the Caribbean islands and the coasts of Central and South America.

Tundra Plovers

American Golden Plover *Pluvialis dominica*

Adult has a mottled dark grey and gold crown, nape and upperparts (though the outer wing is dark grey). The face is black, the remaining head and the sides of the neck white. The underparts are black. Females are as males, but with white streaking on the flanks. Winter adults lose the bold patterning and colours and are grey-brown with some gold mottling.

American Golden Plovers breed in western North America from Alaska to Hudson Bay, and also on Banks, Victoria, Southampton, and western Baffin Islands. Has also bred on Wrangel Island but is not established there. Wintering birds occur primarily in the southern United States and Central America, but they fly as far south as Argentina.

Eurasian Golden Plover *Pluvialis apricaria*

Very similar to American Golden Plover though slightly larger, and has white flanks stretching from the white facial pattern to white undertail-coverts. Winter adults are pale golden brown, quite different from American Golden Plover. Eurasian Golden Plovers are more likely to be seen away from the shore than most other plovers. Highly territorial when feeding, chasing other shorebirds away from its chosen patch.

Eurasian Golden Plovers breed in north-east Greenland (in small numbers), on Iceland, in Scandinavia and northern Russia as far east as the southern Taimyr Peninsula. Absent from Svalbard (though breeding records do exist) and the western Arctic islands of Russia. Greenland and Icelandic birds winter in Iberia, Scandinavian and Russian birds in Iberia and North Africa.

Pacific Golden Plover *Pluvialis fulva*

Very similar to Eurasian Golden Plover, but with longer, darker grey legs, a uniform grey-brown underwing, and less white (and less well-defined white areas) on the flanks. The upperparts are darker overall and less heavily mottled with gold. In winter the birds lose the distinctive pattern on the underparts, being grey, smudged brown.

Pacific Golden Plovers breed in Russia east of the Yamal Peninsula and in western Alaska (though they are uncommon there). Breeds to Russia's north coast, but not on any Arctic islands. In winter the birds move as far south as Australasia.

Grey Plover (Black-bellied Plover) *Pluvialis squatarola*

Adults are patterned as the European Golden Plover, but with very different coloration, having white/pale grey upperparts which lack any gold mottling, and with the black on the underparts extending to the rear flanks. Females are less boldly marked. In winter the upperparts are more subdued, the underparts white, mottled with brown spotting.

Grey Plovers breed in Greenland (but are rare there), in Russia from the White Sea to Chukotka as far north as the coast, and on the New Siberian Islands and Wrangel Island. Breeds discontinuously in North America, being found on the west and south-west coasts of Alaska, but rarely in the north, in Yukon and western NWT. But also breeds on Banks, Victoria and Southampton Islands and on the nearby mainland, and on western Baffin Island. Eurasian birds winter in south-west Europe, south-east Asia and Australia. North American birds move to the Atlantic and Pacific coasts of the southern United States, and into Central and South America.

a Eurasian Golden Plover in breeding plumage.

b American Golden Plover in breeding plumage.

c Pacific Golden Plover in breeding plumage.

d Grey Plover in breeding plumage.

e Eurasian Golden Plover in winter plumage. The juvenile plumage of the species is very similar.

f Pacific Golden Plover in winter plumage. The species' winter plumage is duller than that of Eurasian birds. American Golden Plovers are similar, but darker on the upperparts. Grey Plovers are closer to the American Goldens in winter, but the upperparts are scalloped grey and white and there is no hint of a golden wash.

Calidris Sandpipers

Calidris waders differ from the plovers in having bills adapted for probing rather than pecking. The bills vary enormously in size and shape, with that of the Spoon-billed Sandpiper being among the most remarkable of any bird. The bills have an array of touch sensors (called Herbst's corpuscles) that allow the birds to 'feel' prey beneath the sand or mud (a characteristic shared with the snipes, Tringa sandpipers, godwits, curlews and dowitchers).

Some calidrids can move the tip independently of the rest of the bill. This ability, termed *rhynchokinesis*, means that the bird can open the bill tip while the rest of the bill stays closed, and so can grab a worm or insect larva that can then be taken by the tongue and consumed without the bill being withdrawn. Together with the amazing touch sensitivity of the bill, this allows the bird to feed quickly and efficiently. Least, Pectoral and Semipalmated sandpipers are known to be able to do this, and others may as well, though this has not been proven. Other long-billed shorebirds – including Eurasian Oystercatcher, Common Snipe and Bar-tailed Godwit – are known to share the ability.

In general Calidris sandpipers have long wings and legs and relatively short tails. The sexes are similar. Most calidrids are northern breeding species and migrate great distances, reaching Tierra del Fuego and Australia: close to a million birds of more than 50 shorebird species overwintering in north-west Australia.

Dunlin *Calidris alpina*

Breeding adults have a black-streaked brown crown, the rest of the head and breast being white with black streaking. The upperparts are brown and black. There is a distinctive black patch on the belly, the rest of the underparts being white. In winter the belly patch is lost and adults become pale grey above and white below. The black belly patch allows Dunlin in breeding plumage to be easily separated from similar shorebirds.

Dunlin breed in east Greenland, Iceland, Jan Mayen, Svalbard, Fennoscandia and across Arctic Russia to Chukotka and Kamchatka, including both islands of Novaya Zemlya, the New Siberian Islands and Wrangel Island. In North America Dunlin breed in west and north Alaska and on the northern Canadian mainland west of Hudson Bay, including Southampton Island. Some Icelandic birds stay on the south coast during the winter, but most Palearctic birds move to western Europe, North Africa and southern Asia. Nearctic birds winter on the east and west coasts of North America and in Central America.

Dunlin are the classic example of speciation due to the uneven distribution of ice during the last glacial maximum, and the existence of refugia, land not covered by ice where species could continue to thrive, formed during the glacial and interglacial periods of the entire 250,000 years of the late Quaternary ice ages. Evidence from studies of mtDNA indicate that identifiable changes occurred each time populations were separated by glaciation events, with the formation of distinct subspecies. The Dunlin is polytypic, with subspecies distinguished only by subtle differences in plumage colour and pattern. mtDNA studies indicate that the oldest form of the bird is the subspecies that breeds in central Canada (*C. a. hudsonia*), and that this split from an ancestral form about 225,000 years ago, a time that coincides with the Holstein interglacial. The studies indicate that further splits occurred during the Emian interglacial (120,000 years ago), and during a glacial period about 75,000 years ago. These data suggest that during interglacial periods the Dunlin was able to expand its range, populations then being isolated by further glaciation. Splits during a glacial period would arise if significant ice tongues developed, isolating populations on ice-free tundra to either side.

a C. a. alpina, Varangerfjord, Norway. This subspecies breeds from northern Norway to the Urals.
b C. a. schinzii, central Iceland. The breeding season is over and the bird is moulting to winter plumage. This subspecies breeds in Greenland, Iceland and the British Isles.
c C. a. kistchinski. This much paler subspecies breeds only in Kamchatka.
d C. a. arcticola, Barrow. This subspecies breeds only in northern Alaska.
e C. a. hudsonia, Southampton Island. This subspecies breeds in Canada, west of Hudson Bay.

The Calidris sandpipers can be difficult to differentiate, a problem most acute with the 'peeps', a group of five small sandpipers in North America. Called peeps because of their calls, the most distinguishing characteristics of the group are given below (for the four largest peeps) and overleaf for the smallest of the five. The first two peeps are distinguished by being larger and having longer wings, as would be expected given their long migratory flights.

Baird's Sandpiper *Calidris bairdii*
Breeding adults have a buff-scalloped dark brown crown, nape and mantle. The supercilium is buff, the rest of the head is buff and white. The upperparts are distinctively dark brown spotted. The underparts are white, heavily streaked buff on the breast. The bill is black, straight and has a fine tip (Photos 129a and b). Winter adults are duller overall and lose the dark spotting of the upperparts. Baird's can be distinguished from other calidrids (apart from White-rumped Sandpiper, which shares the trait) by the wing-tips, which project beyond the tail. The wings describe an oval as the bird walks and pecks: Baird's also nod their heads almost constantly. Winter adults are more subdued.

Breeds in Chukotka and on Wrangel Island, and across North America from northern Alaska to Baffin Island. On the Canadian mainland breeds only to the west of Hudson Bay; also breeds in north-west Greenland. Wintering birds fly to South America, mainly the west coast as far south as southern Chile and Tierra del Fuego.

White-rumped Sandpiper *Calidris fuscicollis*
Breeding adults are very similar to Baird's Sandpiper, but the base head colour is white and the streaking is dark brown. The breast is more heavily streaked, the upperparts more rufous, and streaked (rather than spotted) black (Photo 129c). The black bill is slightly decurved. In flight the bird is easily distinguished from Baird's by the highly visible white rump. The birds are found in a much broader range of habitats than Baird's, which prefers a drier habitat. In winter the birds are dull grey with some brown streaking.

Breeds in Canada from the Yukon to western Hudson Bay and on the southern Arctic islands; rare in Alaska and east of Hudson Bay. In winter the birds head for the west coast of southern South America, flying as far as Tierra del Fuego and the Falkland Islands.

Semipalmated Sandpiper *Calidris pusilla*
Very similar to Baird's Sandpiper, but smaller and without the distinctive black spotting or the long wings. Breeding adults have pale grey-white heads, the crown and nape heavily streaked dark brown and grey. The breast and flanks are white streaked with brown, the remaining underparts white. The upperparts are dark brown scalloped with pale grey and chestnut. The straight bill is black, blunt-tipped and broader than in the two larger peeps (Photo 129e). In winter pale grey above, white below, retaining a grey-brown streaked breast band.

Breeds across northern North America, including the southern Arctic islands. In winter seen on Caribbean islands and on the coasts of South America as far south as Uruguay.

Western Sandpiper *Calidris mauri*
Very similar to Semipalmated Sandpiper, but head and upperparts are distinctly chestnut and there is darker, heavier spotting on the underparts. The spotting is lighter on the belly, but that again differentiates the bird from Semipalmated, on which there is no spotting in that area. The black bill is slightly decurved (Photo 129f). In winter adult is pale grey above, white below. Western is almost as pale as Sanderling in winter, but retains a collar of pale grey-brown streaks.

Breeds in Chukotka and western Alaska. In winter Russian birds cross the Bering Sea to join Alaskan birds on flights south to California and the Caribbean islands.

d This rather poor photograph (the day was misty, the light poor, and the author's skills somewhat lacking) well illustrates the longer wing, extending beyond the tail, of the two larger peeps. But the bill, though Baird's-like (as was the plumage) is too long. Dunlin and Baird's Sandpiper are known to hybridise, and it seems likely this is a hybrid.

Least Sandpiper *Calidris minutilla*

The smallest of the peeps: Indeed, the smallest of the world's shorebirds. Breeding adults are very similar to Semipalmated Sandpipers, but have yellow or yellow-green legs, rather than black. In winter the bird is grey above with significant brown scalloping. The breast is grey-brown, the rest of the underparts white.

Least Sandpiper breeds from western Alaska to Labrador, but not on the Canadian Arctic islands. In winter the birds fly to the southern US states, Caribbean islands, and Central and northern South America.

Curlew Sandpiper *Calidris ferruginea*

Status: IUCN: Near threatened (population decline due to habitat loss along migration route and at wintering grounds in east Asia/Australia).

Breeding adults are attractive birds with a black-streaked white crown, the rest of the head, breast and belly being chestnut brown. The upperparts are dappled chestnut, black and white. In winter the colours are lost, the birds being pale grey above and white below. Male Curlew Sandpipers stay on the breeding grounds for just a couple of weeks; they spend their entire time there attempting in copulate with as many females as they can before departing.

Curlew Sandpipers breed from the Taimyr Peninsula to Chukotka, and probably also on the New Siberian Islands. In winter the birds move to Africa, southern Asia, Indonesia and Australia.

Red Knot *Calidris canutus*

Status: IUCN: Near threatened (population decline due to habitat loss along migration route and at wintering grounds primarily in east Asia/Australia).

Very similar in pattern and colour to the Curlew Sandpiper, though the underparts of Red Knot vary from red-brown to salmon pink, while the upperparts are dappled red-brown, black and grey, and the patterning is less bold. However, the main diagnostic is the Red Knot's short, straight bill which is very different to the longer, downcurved bill of the Curlew Sandpiper. The legs and feet are also dark olive, whereas those of Curlew Sandpiper are black.

Red Knots breed from Russia's Taimyr Peninsula, the New Siberian Islands and Wrangel, and inland Chukotka. Breeds in northern Alaska (but not consistently), Canada's Melville Peninsula and many Arctic islands; also breeds in north and north-east Greenland. However, this distribution is patchy, the birds often being very local. In winter, Greenlandic and some east Canadian birds move through Iceland to western Europe. North American birds move to the coasts of the southern United States, to Central America and as far south as Tierra del Fuego. Central Russian birds move to Africa, while east Russian birds fly as far as Australia.

The Red Knot was once one of the most abundant of all North American waders, but relentless hunting during its annual migration severely reduced numbers. Thankfully the hunting has now diminished. Red Knot have exceptionally long migration flights; it has been estimated that during its migrations an individual bird may fly up to 30,000km. Some birds cross the Atlantic in one flight, accumulating subcutaneous fat before the flight that can total as much as 80% of the bird's body weight. As with the Curlew Sandpiper, the principal reason for concerns over population decline is the loss of habitat near the Yellow Sea for migrating birds.

a Curlew Sandpiper in breeding plumage.
b Curlew Sandpiper in winter plumage.
c Flock of Red Knot in breeding plumage.
d Least Sandpiper.
e Red Knot.

b
e

Pectoral Sandpiper *Calidris melanotos*

Breeding adults have dark brown crowns streaked black and brown, and a vague white supercilium. The throat is white, heavily streaked dark brown, the streaking continuing on the buff neck and breast. However, the coloration ceases abruptly and so forms the pectoral band of the name. Beyond the band the underparts are white. The upperparts are dark brown, scalloped chestnut and buff. Wintering adults maintain the breeding pattern but are duller and browner overall. Adult males are larger than females and tend to have darker breasts. Males have a curious hooting call amplified by the expansion of an air-filled sac, which causes the neck/upper breast to puff out. This strange 'wow–wow' call carries over large distances; it can be particularly eerie on misty days.

Breeds from the Taimyr Peninsula to Chukotka, to the north coast but absent from all islands except Wrangel. In North America breeds from north-western Alaska along the mainland coast to Hudson Bay and on the southern Arctic islands (as far north as southern Ellesmere Island). In winter all birds move to South America, with small numbers in Japan and Korea.

Sharp-tailed Sandpiper *Calidris acuminata*

Very similar to the Pectoral Sandpiper and ranges overlap, but Sharp-tailed in breeding plumage has brighter upperparts and much more extensive spotting on the underparts so there is no clear pectoral band. Also has a distinct white eye-ring visible at close quarters. In winter the birds are duller, and virtually indistinguishable from Pectoral Sandpipers.

Breeds in northern Yakutia between Lena and Kolyma rivers. In winter the birds fly to Australasia.

Purple Sandpiper *Calidris maritima*

Breeding adults are the darkest of the calidrids. The head is buff streaked dark brown, the underparts pale grey but heavily spotted dark brown, with the spotting thinning towards the vent. The upperparts are chestnut and white, scalloped dark brown. In winter adults are black-scalloped slate grey above, grey-streaked white below. Purple Sandpipers tend to eat more vegetation than other calidrids, particularly algae. The birds feed very close to the waterline and are occasionally collected by tides or waves.

Breeds on Iceland, Jan Mayen, northern Fennoscandia, Svalbard, Franz Josef Land, Novaya Zemlya, the Taimyr Peninsula and the southern island of Severnaya Zemlya. In North America the bird breeds on Baffin and Southampton Islands, and southern Ellesmere Island. Also breeds in southern Greenland. In winter many Icelandic birds are resident. Greenland birds move to Iceland and the British Isles, while birds from Svalbard and Russia to the Norwegian coast or western Europe. Canadian birds move to coasts of north-east United States.

Rock Sandpiper *Calidris ptilocnemis*

Very similar to Purple Sandpiper (indeed, so similar that Aleutian birds are almost identical and some authorities consider the two taxa conspecific). Elsewhere the range of the species is the best way of distinguishing the two, though outside the Aleutians Rock Sandpipers tend to be paler and brighter. Winter adults are as Purple Sandpiper, though non-Aleutian birds tend to be paler.

Breeds in eastern Chukotka and on the Commander and Kuril islands, in western Alaska, the Pribilof and Aleutian islands. In winter the birds move to the east coast of the United States as far south as California.

a Pectoral Sandpiper in breeding plumage.

b Nesting female Sharp-tailed Sandpiper. The white eye ring can be seen in this photograph.

c Sharp-tailed Sandpiper in breeding plumage. Both shots of the Sharp-tailed Sandpipers were taken on the Taimyr Peninsula.

d Purple Sandpiper in breeding plumage, north-east Greenland.

e Rock Sandpiper C. p. couesi, the Aleutian subspecies.

Sanderling *Calidris alba*

Breeding adults have a black-streaked rufous head, neck and breast, the rest of the underparts white. The upperparts are black, white and rufous, dappled grey. In winter the birds are pale grey above, white below and are much paler than other calidrids.

Breeds in northern Greenland, Svalbard, the Taimyr Peninsula, Severnaya Zemlya and the New Siberian Islands, and on the eastern Canadian Arctic islands north to Ellesmere Island. In winter Greenlandic birds move to west Africa, Asian birds go to Australia and southern Africa, and Canadian birds fly to both coasts of the United States and to Central and South America (as far as south as Tierra del Fuego).

Little Stint *Calidris minuta*

The smallest Palearctic calidrid. Breeding adults have a rufous head finely streaked with black. The throat is white blending to a buff breast, speckled with dark brown spots. The remaining underparts are white. The upperparts are chestnut with black scalloping, though the flight feathers are dark grey-brown. In winter adults are grey or grey-brown above and on the breast, the rest of the underparts are white.

Little Stints breed in northern Fennoscandia and in Russia from Cheshskaya Guba to Chukotka, including the southern island of Novaya Zemlya and the New Siberian Islands. In winter the birds move to North Africa, the Middle East and southern Asia.

Long-toed Stint *Calidris subminuta*

Very similar to both the Little Stint (which is brighter and more rufous overall and has black rather than yellow-green legs) and Least Sandpiper (though the ranges do not overlap). The toe of the name is the middle one (though the hind toe is also longer than in other calidrids), but the value of this difference in the field is somewhat limited. Breeding adults have a white head and breast, each finely marked with brown and dark brown. The remaining underparts are white. The upperparts are grey-brown with darker spots and streaks. In flight the long toes project beyond the tail, but this can be difficult to observe in a fast-flying bird. In winter adults are much duller.

Long-toed Stints are usually found on upland tundra, and are rare. They breed in isolated pockets in central Siberia east of the Ob River, particularly near Magadan, in southern Chukotka, northern Kamchatka and on the Commander Islands. Wintering birds move to south-east Asia and Australia, where they frequently form flocks with other stints.

Temminck's Stint *Calidris temminckii*

Similar to Little Stints, but much more subdued and with yellow-green legs and feet. Breeding adults have grey-brown heads and breasts (but a white throat) finely streaked dark brown. The rest of the underparts are white. The upperparts are brown dappled with black spots. In winter the birds have the same colour pattern but are dull grey-brown.

Temminck's Stints breed from Fennoscandia to Chukotka, to the north coast, though absent from Russia's Arctic islands apart from the New Siberian Islands. In winter European birds move to Mediterranean coasts, while Asian birds head for southern Asia and Japan.

Red-necked Stint *Calidris ruficollis*

Status: IUCN: Near threatened (population decline due to habitat loss along migration route, particularly in area near Yellow Sea).

Breeding adults have orange-red heads and upper breast, the crown streaked darker. The remaining underparts are white, with some dark brown streaking on the breast below the colour patch. The upperparts are rufous or orange-red with black-and-white scalloping. In winter adults are grey on the upperparts and breast, with white underparts.

Breeds patchily from Russia's Taimyr Peninsula to Chukotka. Also breeds sporadically in western Alaska. Migrating birds fly to southern China, Indonesia, Australia and New Zealand.

Sanderling in 'a' breeding plumage and 'b' winter plumage.
c Little Stint.
d Long-toed Stint. The long middle toe is prominent in this photograph.
e Red-necked Stints.
f, g Temminck's Stints.

b

d

f

g

Other Sandpipers

Buff-breasted Sandpiper *Tryngites subruficollis*

Status: IUCN: Near threatened (overhunting in the 19th century caused a dramatic decline from which the species has not recovered. The exact reasons for this are unknown, but may be associated with loss of wintering habitat and exposure to pollutants).

Attractive, but rare, birds. Breeding adults have a buff head, neck and breast, with paler buff belly and vent. The crown and forehead are streaked brown and there is often some streaking on the flanks. The upperparts are buff, more heavily streaked with black and dark brown. In winter adults retain the buff base colour, but the streaking becomes less heavy and more subdued – brown rather than black. The birds differ from most other shorebirds in being usually found in dry areas, and in pecking rather than probing for their food.

Breeds on Wrangel Island, Aion Island and near Cape Yakan in north-western Chukotka, in northern Alaska and the Yukon, and on southern Canadian Arctic islands. In winter both Russian and American birds fly south to Argentina and Paraguay.

Spoon-billed Sandpiper *Eurynorhynchus pygmaeus*

Status: IUCN: Critically endangered (an initially small population is in rapid decline due to low breeding success rate, coupled with habitat loss in breeding, migration and wintering areas as well as hunting and climate change).

Perhaps the most remarkable of all Arctic breeding species. It is, however, also one of the rarest, with a population that might be as low as 1500. In an effort to save the species from what appeared to be inevitable extinction, projects have been set up involving Russian ornithologists at the breeding grounds, conservationists in Bangladesh and Myanmar (to reduce hunting of wintering birds) and the Wildlife and Wetland Trust (WWT) at Slimbridge in Gloucestershire, UK. The latter have taken eggs from Chukotka and are rearing birds with minimal human contact in a simulated natural environment in the hope of being able to introduce breeding adults back into the wild.

The most extraordinary feature of the species is the bill, which is broad at the base, then tapers before flattening into a diamond shape. Breeding adults have a chestnut head, the forehead, crown and nape being streaked dark brown. The breast is smudged chestnut. The upperparts are mottled dark brown, chestnut and white. Winter adults are much paler, the chestnut coloration becoming buff or even white, the streaking grey-brown. The bill is so unusual and so particular a shape that it is diagnostic if it can be clearly seen, but with less-than-perfect views the bird can be confused with Red-necked Stint.

Spoon-billed Sandpiper breeds only in far eastern Chukotka and northern Kamchatka. Migrating birds head through Korea, Japan and China to wintering grounds in the Indian sub-continent.

Stilt Sandpiper *Micropalama himantopus*

Breeding adult has a white head, the crown streaked chestnut. The supercilium is white and very distinct, and the ear-coverts are bright chestnut. The rest of the face and the neck is white, streaked brown. The underparts are white, heavily streaked and barred with dark brown. The upperparts are white, heavily blotched dark brown. Winter adults are much as breeding adults, but much more subdued, lacking the chestnut patches on the head, and with upperparts that are much duller – a uniform pale grey-brown.

Stilt Sandpiper breeds from northern Alaska (where it is rare) to Bathurst Inlet and on southern Victoria Island. Also breeds at the south-west corner of Hudson Bay. In winter the birds move to central South America.

a Buff-breasted Sandpiper in breeding plumage.
b Buff-breasted Sandpiper in winter plumage.
c An historically interesting photograph – the first taken of a Buff-breasted Sandpiper nesting in Russia. The sandpipers were first found breeding on Wrangel Island in 1979, but the first photograph was not taken until 1981.
d Female Spoon-billed Sandpiper with brood.
e, g Spoon-billed Sandpipers.
f Stilt Sandpiper in breeding plumage.

Broad-billed Sandpiper *Limicola falcinellus*

Rare and elusive, though the range size means it is not considered to be vulnerable despite it being estimated that the population is in decline. Recognisable for the bill, which is broad at the base (when viewed from above) and tapers towards the downcurved tip. Breeding adults have dark brown or black crowns delineated by a thin white stripe, which meets the white supercilium in front of the eye. The rest of the head and neck (apart from a white throat) is white with black streaking. The underparts are white, with significant black streaking on the breast, and less heavy streaking on the flanks. The streaking takes the form of arrowheads, these occasionally distinct, but often merging to form long streaks. The upperparts are grey-brown with heavy black markings. Wintering birds lose the dark crown, the head and upperparts being white or pale grey with black and grey-brown mottling.

Broad-billed Sandpipers breed in northern Fennoscandia and in isolated areas of Asian Russia as far east as the Kolyma river. In winter the birds move to southern Africa, southern parts of Arabia and India, and Indonesia and northern Australia.

Ruff *Philomachus pugnax*

Exhibits extreme sexual dimorphism. Breeding males (ruffs) have a bare red or orange-red face with many warts, these usually red, but occasionally yellow or green. The prominent ear tufts, which cover the sides of the nape, and the circular ruff, which extends to the breast, are highly variable in colour, chiefly being white, black or shades of rufous brown. The ruff may also have complete black rings, incomplete rings or no rings. The underparts are white, the breast heavily streaked black. The upperparts are brown, buff and black. Breeding females (reeves) are much smaller and duller, lacking the bare face, ear coverts and ruff. The head is grey-brown, streaked darker. The upperparts are grey-brown, spotted with black. The underparts are white with heavy buff and black streaking on the breast and flanks. In winter male birds lose all the showy breeding plumage and become rather anonymous grey-brown birds. Wintering females lose the black feather-centres that give the black spotting, and so are as breeding, but duller.

Breeding male Ruffs have elaborate lek behaviour, the lek arenas being traditional sites that may have been used over hundreds of years. At these sites the birds form three distinct groups – residents, which hold territories at the arena; migrants, which attempt to obtain territories, and may hold territories at other arenas; and opportunistic satellites. In general the bird that holds the central territory is the dominant male who will have fought off challengers for right of possession; in general, females will head for him, ignoring the displays of other males they pass. Studies show that the colour variations of male ear-coverts and ruff are associated with the groupings. Residents invariably have black ear-coverts, satellites white ear-coverts and ruffs: satellites are tolerated by residents as this coloration is less attractive to females. Satellites will hold a territory in the absence of the resident bird, re-assuming a satellite role when the resident returns. Some authorities suggest satellites display a form of behavioural dimorphism and that this and white coloration are heritable.

The displaying ruff flutters his wings and jumps in the air, with the ruff expanded to its full circle and the ear-coverts raised. The ruff will mate with any interested reeve, so polygamy is common. However, some studies show that monogamy is also practiced. Display and copulation are the ruff's only contribution to the next generation, the reeve incubating the eggs and caring for the chicks. As if to finally confirm female prejudices regarding the character of males, the testes of the ruff do indeed weigh more than his brain. It is, though, worth noting that reeves frequently mate with several ruffs in the arena, studies suggesting that more than half the broods contain chicks with different fathers.

Ruffs breeds in Fennoscandia and across Russia to the border of Chukotka. Breeds to the north coast in places, but is absent from all of Russia's Arctic islands. Wintering birds occur in western Europe, along the coast of the Mediterranean, and as far south as the coasts of India and southern Africa.

a Incubating female Broad-billed Sandpiper.
b Adult Broad-billed Sandpiper in breeding plumage.
c, e and f Male Ruffs in breeding plumage.
d Ruffs and reeves at a lek.

Spotted Redshank *Tringa erythropus*

With their black heads, necks, and underparts adult Spotted Redshanks are very distinctive. The upperparts are black with white spotting, apart from the back and upper rump, which are white. The legs are black; the bill is black with a red base. Wintering birds are grey above and white or pale grey below, the legs being red. Wintering birds can be difficult to distinguish from Common Redshanks as, although bigger, the size difference is slight. In general, wintering Spotteds are paler and a more uniform grey.

Females spend a very short time at the breeding grounds. They chose a mate, copulate and lay, and, after a few days of incubation they depart for the wintering grounds, leaving the male to take over incubation and chick care.

Spotted Redshanks breed in northern Fennoscandia, on the eastern shores of the White Sea, and from the Ob estuary east to Chukotka, north towards the coast, except on the Taimyr Peninsula where it breeds only in the south. In winter the birds move to the Mediterranean coast, central Africa, and the coasts of India and south-east Asia.

Common Redshank *Tringa totanus*

Breeding adults have a pale olive head and breast, finely streaked darker. The rest of underparts are white with dark spots. The nape, scapulars and inner wing are olive with darker streaking, the rest of the upper body being white. The tail has dark barring. The bill is black with an orange base. In winter adults lose the distinct streaking and are dull mottled grey-brown. The red legs are diagnostic as regards species other than the Spotted Redshank.

Common Redshanks are highly vocal when intruders encroach on their territories, the sharp whistling call bringing adjacent birds quickly to the scene, so that the intruder is rapidly surrounded by birds and bombarded with noise. The intruder (perhaps a potential predator) is soon driven away by the incessant racket. A loose colonial nesting system has resulted from this efficient anti-predator strategy and other species occasionally nest close to the Redshanks to take advantage.

Common Redshanks are essentially subarctic but breed on Iceland and western Fennoscandia. Icelandic birds winter on the British Isles and mainland Europe.

Lesser Yellowlegs *Tringa flavipes*

A long-legged, elegant bird. Breeding adults have white or pale grey heads, finely streaked dark brown. The breast is pale grey, heavily marked darker grey, the rest of the underparts are pale greyish white, the marking diminishing towards the vent. The upperparts are dark grey-brown spotted with white. The thin bill is black. The legs are bright yellow. In winter adults are as breeding, but overall more subdued. In areas where they overlap it can be difficult to distinguish Lesser from Greater Yellowlegs, though the calls differ. When seen together the size difference is clear. However, Greater Yellowlegs are essentially subarctic and may only be seen together with their Lesser cousins in southern Alaska and southern parts of western Canada. The latter includes Churchill, where the two species may be seen together.

Lesser Yellowlegs breeds in central and southern Alaska (rare on the west and north coasts), around the Mackenzie delta, then more southerly to the southern shores of Hudson Bay. In winter the birds move to the coasts of the southern United States, to Central America, the Caribbean islands, and South America as far south as Tierra del Fuego.

a, b Spotted Redshanks in breeding plumage.
c, d Common Redshanks in breeding plumage. Photo 'c' shows a characteristic pose of the species, which usually raises its wings in this fashion after landing. The raised wing pose is also used against conspecifics and intruders during breeding.
e, f Lesser Yellowlegs in breeding plumage.

Wood Sandpiper *Tringa glareola*

Breeding adults have a pale grey head with a dark brown crown, streaked pale grey. The neck and underparts are pale grey, heavily streaked dark brown except for the belly and vent. The upperparts are dark brown, heavily barred and spotted pale grey-brown. Winter adults are as breeding, but the underparts are more subtly streaked and the upperparts are grey-brown.

Wood Sandpipers are a boreal species, found at the timberline but rarely on the tundra. Breeds in Fennoscandia, across Russia to central Chukotka and throughout Kamchatka. In winter the birds are seen in southern Africa, India and Australasia.

Solitary Sandpiper *Tringa solitaria*

Breeding adults have olive-brown heads, finely streaked white. There is a white eye-ring. The throat and underparts are greyish white, the breast heavily streaked dark brown. The upperparts are dark brownish-black with some white or pale buff spotting. In winter adults are patterned as breeding, but grey-brown with little streaking. Much less common than the Spotted Sandpiper, with which it might be confused. The orange legs and bill of Spotted are diagnostic, with Solitary Sandpiper having a dark grey bill and yellow-green legs.

Solitary Sandpipers breed in central and southern Alaska (where it is uncommon, and rare on the west and north coasts), on the Mackenzie delta, then more southerly to the southern shores of Hudson and James Bays and into southern Quebec and Labrador. In winter the birds are seen in eastern Central America, on Caribbean islands and in northern South America.

Spotted Sandpiper *Actitis macularius*

Breeding adults have greenish-brown heads and napes. The throat, front neck and underparts are white with heavy black spotting as far as the breast, then less heavy spotting on the belly, flanks and vent. The upperparts are greenish-brown. In winter adults lose the heavy spotting on the underparts, and the upperparts are brown. The bill changes from bright orange with a black tip to dark brown in winter.

Spotted Sandpipers breed in central and southern Alaska (but are rare on west and north coasts), on the Mackenzie delta, the southern shores of Hudson Bay and James Bay, and in central Quebec and Labrador. In winter the birds fly to south-west coastal United States, to Central America and the Caribbean islands, and to northern South America.

Common Sandpiper *Actitis hypoleucos*

Adults have grey-brown heads and breasts with darker eye stripes. The rest of the underparts are white. The upperparts are bronze-brown, marked by dark arrowheads. In flight a white wing-bar is distinctive. The long tail and persistent bobbing of the rear body are diagnostic.

Common Sandpipers are essentially subarctic, but do breed in northern Fennoscandia, more southerly across Russia, and throughout Kamchatka. In winter the birds move to Africa, the Middle East and southern Asia.

a Wood Sandpiper in breeding plumage.
b Spotted Sandpiper in breeding plumage.
c, d Solitary Sandpipers in breeding plumage.
e, f Common Sandpiper in breeding plumage.

c

d

f

Snipes and Dowitchers

These shorebirds are all generally stocky birds with long bills that share feeding characteristics. In Snipe the sexes are similar, and summer and winter plumages are identical. In the dowitchers, the sexes are similar, but summer and winter plumages differ.

Snipes are among the few birds that deliberately make a non-vocal sound. During display flights the birds dive at an angle of c.45° while fanning the tail. The two outer feathers of the tail are stiff and have asymmetric vanes, the leading edges being narrow. The bird holds these feathers at right-angles to the body; at speeds of about 60km/h or higher they vibrate. The air through the vibrating feathers is modulated, creating a loud drumming noise usually called winnowing, which can be heard over considerable distances. It is often heard before the observer has seen the bird. Although it is male snipe that are chiefly responsible for the drumming, females also make the sound, particularly early in the breeding season.

Common Snipe *Gallinago gallinago*
Adults have dark brown crowns with a thin central buff stripe. The rest of the head is buff with darker stripes through the eye, the buff above the eye appearing as a broad supercilium. The upperparts are mottled dark and light brown. The breast is buff, streaked dark brown. Underparts are white, the flanks streaked dark brown (Photos 145a and c).

Breeds in Iceland, northern Fennoscandia, and across Russia, though rarely to the north coast and not on Arctic islands. Icelandic birds are resident, but European birds move to the Mediterranean coasts and a sub-Saharan belt; Asian birds fly to India and south-east Asia.

Wilson's Snipe *Gallinago delicata*
Until recently considered a Nearctic subspecies of Common Snipe. The plumage is essentially identical, though the underwing tends to be darker and there is no, or only a very thin, white trailing edge on the upperwing (Photo 145b). However, Wilson's has 16 tail feathers rather than the 14 of the Common, and the outer feathers are shorter.

Breeds across North America from Alaska to Labrador, though only to the north coast near the Mackenzie delta. Absent from islands of the Canadian Arctic. In winter the birds move to the southern United States, Central America and northern South America.

Long-billed Dowitcher *Limnodromus scolopaceus*
Adults have a chestnut head with a paler supercilium, and a dark brown crown. Underparts are chestnut or pale chestnut, the neck and upper breast finely streaked and barred dark brown. The upperparts are mottled dark brown, chestnut and buff. In winter, adults are pale grey-brown above. The breast is mottled grey-brown, the rest of the underparts white. When feeding the birds probe very quickly, sewing machine-like. They often feed in shallow water, so for the photographer, getting a shot with the long bill exposed rather than half-buried is tricky (Photo 145d).

Breeds in Russia east from the Yana River and south to Anadyr, and on Wrangel Island. Breeds patchily in North America eastwards to the west and south shores of Hudson Bay. Also breeds on southern Victoria Island. Both Russian and American birds winter in the south-west United States and in Mexico.

Short-billed Dowitcher *Limnodromus griseus*
Until the 1950s the two northern species of dowitchers were considered conspecific, largely because of the similarity of the Short-billed subspecies that overlaps in range with the Long-billed near southern Hudson Bay. The plumage of this race, *L. g. hendersoni*, the 'Prairie' Short-billed, is extremely similar to the Long-billed, though the upperparts are mottled gold and black, and the underparts are orange rather than chestnut (Photo 145e). The 'Pacific' Short-billed, *L. g. caurinus*, overlaps with the Long-billed in southern Alaska, but is usually easily distinguished, having a white belly and vent with heavy dark brown barring. The longer bill and legs of the Long-billed are diagnostic if the birds are viewed together.

Breeds in three distinct areas of North America – southern Alaska, around the southern shores of Hudson Bay and in southern Quebec and Labrador. In winter the birds move to the east and west coasts of the southern United States, Central America, the Caribbean islands, and the east and west coasts of northern South America.

Godwits and Curlews

While these are two distinct groups, they are all elegant, long-legged, long-billed shorebirds. The curlews are brownish and have down-curved bills, while the godwits are usually more colourful, their breeding plumage red-washed, and have straighter bills. The origin of the name 'godwit' is debated, some claiming it derives from the Old English *god whit* – 'good creature' because it offered fine eating, while others suggest it is an onomatopoeic rendering of the call of the birds. In support of the latter contention, the name curlew is a likely onomatopoeic rendering of the species' call.

Bar-tailed Godwit *Limosa lapponica*

Status: IUCN: Near threatened (the species comprises four (or five, opinions differ) subspecies. The nominate breeds in northern Scandinavia and eastwards to the Taimyr Peninsula and migrates to western Europe and Africa: its population appears stable. Further subspecies breed in eastern Siberia and on Alaska's western and northern coasts. One of these subspecies (*L. l. taymyrensis*) migrates to west/south-west Africa: little is known about the population trend of this group. The remaining Siberian/Alaskan birds travel across east Asia to Australasia. The population of this group is declining rapidly because of severe habitat loss, particularly around the Yellow Sea).

Breeding males are handsome birds with rufous brown heads and underparts, the crown streaked black-and-white. The bill is long and slightly upcurved. The distal half is dark brown or black, the basal half pinkish-yellow. The upperparts are dark brown, chestnut and white. The tail is barred dark brown and white. Breeding females are similar but paler overall, with a white belly and vent. The female's bill is longer. In winter adults are as female, but pale grey-brown rather than pale chestnut.

For distribution see above. Recently, telemetry has confirmed that one bird flew non-stop from New Zealand to China's Yellow Sea, a distance of a little over 11,000km, at an average speed of 51km/h. Another tagged bird (of subspecies *L. l. baueri*) was later found to have flown non-stop from breeding grounds in Alaska to New Zealand, a distance of 11,680km, the longest known flight of any bird.

Black-tailed Godwit *Limosa limosa*

Status: IUCN: Near threatened (populations in some parts of the species' extensive range are declining rapidly due to habitat loss).

Breeding males are similar to Bar-tailed Godwits but the breast is (usually pale) rufous, the rest of the underparts white, with heavy black barring on the flanks. The bill tip is dark brown or black, the remainder orange-yellow. The tail is black. The sexes are similar. In flight the feet extend well beyond the tail. In winter the upperparts and breast are pale grey-brown.

Black-tailed Godwits breed on Iceland and patchily on the Norwegian coast, but are otherwise more southerly in their Palearctic distribution.

Hudsonian Godwit *Limosa haemastica*

Breeding males have pale grey heads and necks, finely streaked dark brown. The crown is dark brown, streaked pale chestnut. The distal third of the bill is dark brown, the rest orange. The underparts are dark chestnut, paler on the vent and undertail, and with dark brown streaking on the breast and flanks. The upperparts are dark brown, chestnut and buff, the rump grey-brown. Breeding females are as males, but with white underparts heavily barred chestnut. In winter adults are grey above, paler below.

Hudsonian Godwit breeds patchily in western North America, at sites on the southern coast of Hudson Bay, near the Mackenzie delta and, rarely, in western and southern Alaska. In winter the birds fly to southern Argentina.

a Flock of Bar-tailed Godwits in winter plumage.

b Bar-tailed Godwit in breeding plumage.

c Black-tailed Godwit in breeding plumage.

d, e Hudsonian Godwits in breeding plumage.

Whimbrel *Numenius phaeopus*

Adults, which are identical in breeding and winter plumage, have grey heads, the crown being dark brown with a grey central stripe. The bill is long and downcurved. The neck and underparts are grey-buff with heavy brown streaking except on the belly and vent. The upperparts are dark brown, white and pale buff. The tail colour varies: it may be pale with some dark barring, or darker without barring (Photos 149a and b).

Whimbrels breed on Iceland, in Fennoscandia and western Russia, then more patchily across Siberia. Breeds in North America in western and central Alaska, but more rarely in the north, on the Mackenzie delta and around the south-western and southern shores of Hudson Bay. In winter Eurasian birds move to sub-Saharan Africa, southern Asia and Australasia, while Nearctic birds move to the southern United States, Central America, the Caribbean islands, and South America as far as southern Chile.

Bristle-thighed Curlew *Numenius tahitiensis*

Status: IUCN: Vulnerable (as a consequence of population decline at wintering grounds due to predation by feral/domestic cats and dogs which hunt the bird when they are flightless during the moult, and human hunters).

Adults are very similar to Whimbrels whose range overlaps. The curlews have pale cream or buff head with a darker crown, this split by a buff stripe. The supercilium is pale buff or cream. The bill is long and downcurved. Neck and underparts are buff or pale cinnamon, neck, breast and flanks heavily streaked brown. The upperparts are dark brown, spotted buff. Breeding and non-breeding plumage identical (Photo 149c). The stiff feathering of the upper leg – the bristled thigh of the name – is diagnostic but of minimal value in the field. The buff, barred tail is diagnostic, as would be the white rump of Palearctic Whimbrels, but Nearctic Whimbrels lack the white rump.

Bristle-thighed Curlew breeds in a few places in western Alaska (the Yukon delta and upland Seward Peninsula). In winter the birds move to Polynesia. Bristle-thighed Curlews are rare, probably numbering less than 5,000. This is strange because in Alaska the available habitat could accommodate many more birds. What limits the population is the size of the winter habitat, this being confined to a handful of islands in the south Pacific. During the winter the birds take invertebrates and rodents, but also feed on gull eggs, occasionally stealing eggs from beneath incubating birds. The stolen eggs are pierced with the bill or broken by being dropped. A bird will sometimes drop a rock on to an egg – one of the few observations of avian tool-use.

Far-eastern Curlew *Numenius madagascariensis*

The largest of the curlews. Adults have buff-brown heads, necks and breasts, each finely streaked with dark brown. The rest of the underparts are paler with less heavy streaking. The upperparts are also buff with dark streaking, the rump paler, but distinctly darker than the white rump of the Eurasian Curlew, which the bird strongly resembles (Photos 149d and e). Winter and summer plumages are similar.

Far-eastern Curlews have a limited range in eastern Russia and north-east China, but some breed around the western and northern shores of the Sea of Okhotsk and on Kamchatka. Wintering birds occur in south-east Asia and Australasia.

Eskimo Curlew *Numenius borealis*

Status: IUCN: Critically endangered (population unknown). CITES: Appendix I.

Overhunting and loss of habitat caused a drastic depletion of this once-numerous species. Migrating birds fed primarily on the Rocky Mountain Grasshopper (*Melanoplus spretus)* that became extinct when the prairies were claimed for agricultural land and fires on remaining areas were suppressed. The present position of the species is unclear. There has been no confirmed sighting since 1963 and no claimed sighting since the early 1980s. Most authorities consider the species to be extinct, noting that a bird with a migration path that took it across the southern USA and on to southern South America can hardly have escaped detection for over 30 years. Even the most optimistic consider the population to be less than 50–100 birds and the long-term survival of the bird seems to be bleak.

c

e

Turnstones, Tattlers and the Surfbird

The two Arctic-breeding Turnstones have bills similar to those of the plovers, being short for pecking at prey rather than probing for it. The name gives away the birds' search method, powerful neck muscles being used to overturn objects. Stone-turning can reveal all sorts of things, alive or dead, the birds consequently having the most varied diet of any shorebirds, as they will eat just about anything that they find underneath.

This section also deals with the Tattlers, one Alaskan, one Asian and the Surfbird, which do not easily fit into the other Shorebird categories. The Surfbird's curious name derives from its feeding habit – close to incoming tides so it is often drenched by surf.

Ruddy Turnstone *Arenaria interpres*
Breeding males are attractive bird with white heads, the crowns finely streaked black. There is a variable pattern of black bands on the head and breast. The rest of the underparts are white. The upperparts are tortoiseshell chestnut/orange and black (Photo 151b). Females have similar pattern, but duller. In winter the pattern remains, but the overall colour is dark grey-brown.

Breeds in north Greenland, Iceland, Svalbard, Fennoscandia and across Russia, including the Novaya Zemlya's south island, New Siberian Islands and Wrangel. Also breeds in west and north-west Alaska, along western Canada's north coast and southern Arctic islands. Icelandic (and Greenlandic?) birds remain on south coast of Iceland in winter. Eurasian birds move to west European coasts, Africa, the Middle East, India, south-east Asia and Australasia. American birds move to the coasts of the southern USA, Central America, the Caribbean islands and South America to northern Chile and Argentina.

Black Turnstone *Arenaria melanocephala*
Breeding adults have black heads with delicate white streaking on the crown and nape, and a prominent white spot at the base of the bill. Upperparts and breast are black, flecked white (Photo 151a). In winter adults are as breeding, but dark brown rather than black, and with little white speckling and no white bill-spot (Photo 151f (right)).

Breeds only in west and south Alaska, and winters on US west coast to Baja California.

Wandering Tattler *Heteroscelus incanus*
The crown, nape and upperparts of breeding adults are uniform slate grey. The cheeks, throat and underparts are white, heavily barred dark grey (Photo 151c). In winter the heavy barring of the underparts is replaced by an overall grey wash.

Breeds in Alaska and the Yukon, but patchily, and is uncommon. Rare in north and central Alaska. In winter the birds fly to south-west coastal USA and across the Pacific Ocean to Hawaii and other remote islands.

Grey-tailed Tattler *Heteroscelus brevipes*
Status: IUCN: Near threatened (population decline due to habitat loss and pollution).
The Tattlers are very similar, and the two occasionally occur together in Alaska. Grey-tailed have paler upperparts, a thin white bar on the upperwing, and the underparts are less heavily marked (Photo 151d), but the differences are subtle, as are differences in calls (the Wandering is trisyllabic, the Grey-tailed bisyllabic, though the alarm calls are identical). The real diagnostic characteristic (which is not as difficult to see as it sounds) is the nasal groove. In Wanderings this is long, around 75% of bill length, but it is much shorter in Grey-taileds.

Breeds on the northern Yenisey river, patchily from the Lena to western Chukotka, and throughout Kamchatka. Some Kamchatka birds are resident, staying close to hot springs, but the majority of Russian birds fly to coastal areas of Indonesia and Australasia.

Surfbird *Aphriza virgata*
Breeding adults have grey and white streaked head, and heavily streaked dark grey-brown breast. The upperparts are gold washed, with black spotting, though the gold fades with ageing (Photo 151e). In winter adults lose the gold upperpart wash and heavy spotting (Photo 151f (left)).

Essentially subarctic, breeding patchily across Alaska, though common only on the south coast. Also breeds in central Yukon, but not common. Winters on west coast of USA.

Phalaropes

Two of the three species of phalaropes (all of which are northern dwellers) are true Arctic breeders and make long migration flights: they are clearly tough little birds, belying their fragile looks. The two Arctic species share many traits, including polyandry. They also share a feeding habit, swimming in tight circles ('spinning') to stir up the water and bring prey to the surface. As well as spinning, phalaropes feed at the shoreline, taking insects found among seaweed and tidal debris. Wintering birds are chiefly pelagic, taking small fish and free-swimming crustaceans, and even picking parasites from the skins of whales.

In all three phalaropes females are larger than males, and have much brighter breeding plumage. In Grey Phalaropes females are about 20% larger, in the Red-necked about 10%. The females court males with aerial and terrestrial displays. After mating and laying a clutch of eggs, the female will abandon the male, leaving him to incubate the eggs and care for the young. The female seeks another male if one is available, laying a second clutch before abandoning him as well. The female stays on the breeding grounds, so if either clutch is lost she can mate and lay again with either male. Such polyandry is uncommon, though not unique, among birds. The chicks are downy and precocial.

Red-necked Phalarope *Phalaropus lobatus*

Breeding females have dark grey crowns and napes, the rest of the neck red-brown. The throat is white. The breast and flanks are grey, the rest of the underparts white. The upperparts are dark grey with buff lines. Breeding male is as female, but duller overall. In winter the sexes are similar, the head white with a dark grey hind crown and nape and prominent black eye stripe, the upperparts pale grey and white, the underparts white with occasional pale grey patching.

Red-necked Phalaropes breed in southern Greenland (but are rare in the north), Iceland and Svalbard, in Fennoscandia, and across northern Russia to Chukotka and Kamchatka, but not on any Russian Arctic islands. Breeds across North America and on Canada's southern Arctic islands. Full details of the species' wintering sites are not known. Asian birds are seen off western South America, in the Arabian Sea and north of Indonesia, but though it is known that the birds congregate in the Bay of Fundy, as many as two million birds being reported, no other Atlantic wintering sites have yet been confirmed.

Grey Phalarope (Red Phalarope) *Phalaropus fulicarius*

The breeding female is a stunning bird with a black crown and white face (apart from a black and dark grey patch at the bill base). The neck and underparts are bright chestnut-red, the upperparts dark brown with cinnamon/buff lines. Breeding males are as females, but the crown is streaked buff and dark grey, the face pattern less sharply defined and overall much duller. The sexes are alike in winter, and are similar to Red-necked Phalarope, though pale grey rather than pale greyish white on the upperparts. The Red-necked also has a much thinner bill. The grey winter plumage is the origin of the European name for the species, North America naming the species for the showier breeding plumage.

Grey Phalaropes breed in north-west Greenland (and also in the north-east, though they are much rarer there), Iceland (where it is among the rarest of breeding birds), Svalbard (where it also very rare) and on the north coast of Asian Russia east of the Yenisey. Also breeds on the New Siberian Islands and Wrangel. Breeds in northern Alaska, around the Mackenzie delta, on Canada's Boothia Peninsula and the north-west shore of Hudson Bay, and on Canada's Arctic islands north to southern Ellesmere Island. The population on Southampton Island has declined drastically in the last few years, for unknown reasons. In winter European birds fly to the Atlantic Ocean off central and southern Africa, while Asian and American birds move to the Pacific Ocean off southern South America.

a Female Grey Phalarope. The breeding plumage belies the European species name.
b, c Red-necked Phalarope pair and mating.
d Red-necked Phalarope in winter plumage. Wintering Greys are similar, but with greyer upperparts.
e Red-necked Phalarope spinning.
f, g A Red Phalarope female approaches a male and solicits mating with a wing-waggling dance. The male seems singularly unimpressed.

Skuas

Of the seven members of the Stercorariidae, four are true Arctic dwellers. Skuas have webbed feet and sharp claws that, combined with a powerful hooked bill, make them formidable predators. They take a variety of foods, not only the fish that might be expected but also birds and eggs, small mammals, and even berries and insects. Both Pomarine and Long-tailed Skuas tend to specialise in hunting rodents and are frequently seen far from the coast and in upland areas: in Norway the Long-tailed Skua is actually known as the Mountain Skua.

The origin of the name 'skua', used in Britain is unclear but it is likely to derive from the Shetland name *skooi* for the Arctic Skua (though this does not help much as the origin of skooi is obscure). One delightful suggestion is that the word has its roots in *skoot*, the Norse word for what would be best termed 'excrement' in polite company, due to the widely held belief that Arctic Skuas consumed the excrement of other seabirds, with the skuas frightening gulls then cleaning up afterwards. In reality, of course, the skuas were frightening the birds into dropping their fish catch as they are aerial pirates, well deserving their American name of Parasitic Jaeger. The word *jaeger* almost certainly derives from the German for hunter, which also fits the birds rather well. The Shetland islanders gave the Great Skua its own name, *bonxie*, which probably comes from the Norse *bunski*, an untidy mess, a word usually applied, in a very definitely non-PC way, to untidy, wizened old women.

Northern skuas are sexually dimorphic, females being larger than the males by 10–15%. The three smaller northern species also exhibit plumage dimorphism, having both light and dark morphs (apart from adult Long-tailed Skuas). →

Great Skua *Stercorarius skua*

Breeding adults are overall cinnamon-brown, the crown, distal section of the flight feathers and tail darker, with buff scalloping on both upper- and underparts. Some adults are paler and intermediate forms are seen. Summer and winter plumages are similar. Great Skuas feed mostly on fish (primarily sand eels Ammodytes sp.), but they are opportunistic and will take birds (species as large as geese), mammals (hares having occasionally been seen taken) and even chicks from adjacent skua nests; some Hebridean birds feed nocturnally on returning storm-petrels.

Breeds on Iceland, Jan Mayen, Svalbard and Bear Island, and on the northern coast of Norway. In winter the birds are pelagic in the North Atlantic from Ireland to central Africa, with some Icelandic birds flying to the waters off Newfoundland. The Great Skua expanded its breeding range in the last half of the 20th century, reaching Svalbard in the mid-1970s. In places the species has ousted the Arctic Skua from breeding sites it once held, with the bigger bird killing both adults and chicks.

Arctic Skua (Parasitic Jaeger) *Stercorarius parasiticus*

The most piratical of the skuas. Breeding adults have two morphs. Pale birds have dark brown crowns/upper heads. The sides of the head and lower nape are yellow, the throat white. The breast has grey-brown sides or a complete collar, the rest of the underparts being white apart from the dark grey-brown undertail. The upperparts are dark slate-grey or grey-brown, apart from the darker flight feathers. Dark-morph adults are overall dark or sooty brown, but with a black crown and paler cheeks. Intermediate forms are also seen. In winter the yellow of the head and nape are lost or duller.

Arctic Skuas breed on the west and north-east coasts of Greenland, on Iceland, Svalbard, Fennoscandia, and across northern Russia to Kamchatka, including Franz Josef Land and Novaya Zemlya. Breeds across northern North America from west Alaska, including Canada's Arctic islands north to southern Ellesmere. In winter the birds move to the coasts of South America, the western and southern coasts of Africa, the Arabian Sea and the coasts of Australasia.

a-c Great Skuas.
d Pale morph Arctic Skua.
e, f Dark, intermediate and pale morph Arctic Skuas.
g Arctic Skua pursuing a Kittiwake in the hope of pirating its prey.

Pomarine Skua (Pomarine Jaeger) *Stercorarius pomarinus*

Two morphs, though dark morphs represent a smaller percentage of the population than in Arctic Skua. Pale-morph adults are patterned as Long-tailed Skuas, but are darker overall and have a much darker underwing. Breeding adults have yellow cheeks and nape. Dark-morph adults are dark brown, usually even darker than dark-morph Arctic Skuas. The central pair of tail feathers of all birds is elongated, twisted at the base and ends in club-like blobs, a diagnostic feature. In winter the yellow of cheeks and nape are lost or duller. Pomarine Skuas primarily feed on rodents and so do not breed where rodents are absent.

It has been suggested that Pomarine Skuas breed on west Greenland, but the lack of lemmings makes this doubtful. Breeds on Svalbard and Russia's northern coast from the White Sea to the Bering Sea. In North America breeds on the west and north coasts of Alaska, patchily on the Canadian mainland and Banks, Victoria and southern Baffin islands. May also breed at remote sites on other islands. In winter the birds fly to the waters of the Caribbean, the Atlantic off northern Africa, the Arabian Sea and the Pacific Ocean off eastern Australia, northern New Zealand, Hawaii and north-western South America.

Long-tailed Skua (Long-tailed Jaeger) *Stercorarius longicaudus*

The smallest skua. Breeding adults have a black upper head/upper nape. The cheeks and lower nape are yellow, the throat white. The breast is white, merging into a pale grey belly, flanks and undertail-coverts. The upperparts are uniformly slate-grey apart from black flight feather and tail. The central tail feathers form long streamers. Although adult birds probably do not show plumage dimorphism, juvenile birds do, with dark-morph juveniles moulting into the 'standard' adult plumage. In winter the yellow cheeks and nape are lost or duller.

Long-tailed Skuas breed in west and north Greenland, on Jan Mayen and Svalbard, in mountainous Scandinavia, northern Fennoscandia and across Russia to Kamchatka. Absent from the Taimyr Peninsula, but breeds on Novaya Zemlya's south island and on Wrangel. Breeds across Alaska and northern Canadian mainland to the western shores of Hudson Bay. Absent from eastern Canada apart from a colony on the east-central shore of Hudson Bay. Breeds on all Canada's Arctic islands to northern Ellesmere. Wintering birds are pelagic off the east and west coasts of southern South America and the west coast of southern Africa.

→All northern skuas are kleptoparasites (pirates) obtaining food by chasing gulls and terns and forcing them to drop or disgorge recent catches, then retrieving the food, usually in mid-air. Arctic Skuas will even migrate with Arctic Terns, so the terns can be pirated along the way. Pomarine and Long-tailed Skuas are piratical in winter, but in summer they tend to hunt their own food. Great Skuas are less frequently piratical. Skuas take long migration flights, these often being overland, hungry birds then incurring the wrath of farmers by attacking chicken or duck flocks. The southern species also migrate over long distances, and vagrants can drift vast distances off course: one South Polar Skua, *Catharacta maccormicki*, a species that breeds on the Antarctic continent, was, sadly, shot near Nuuk, Greenland.

Skua mating displays are limited, amounting to little more than the male adopting an upright posture in front of the chosen mate, though the bond is reinforced by courtship feeding. The pair bond is monogamous and life-long or, at least, long-lived. Two eggs are laid in a rudimentary nest. These are incubated by both birds (though mainly the female in Great Skuas). The chicks are semi-altricial and are fed by regurgitation. Skuas are long-lived and mature slowly, Great Skuas not breeding until they are seven or eight years old. The parent birds are highly aggressive in defence of their eggs and young, dive-bombing intruders to scare them off. Such attacks are frequently not bluffs: both Per Michelsen and myself have been knocked down by Great Skuas and had blood drawn by Long-tailed Skuas, to prove the point. Interestingly, the smaller skuas will also adopt the plover technique of luring a predator away by feigning injury, though this approach is used much less often than a direct attack.

a-c Pomarine Skuas.
d-g Long-tailed Skuas. On Ellesmere Island a pair of Long-tailed Skuas were content to share breakfast with Per Michelsen and Richard Sale, but attacked them ferociously if they crossed an invisible line that marked their territory. To hunt rodents the birds regularly used Michelsen (the taller of the two men) as a look-out.

c

e

g

Gulls

Gulls represent not only one of the most readily identifiable groups of birds for the casual observer (even if the different species are often more difficult to distinguish), but one of the most successful of bird families, in part due to their being supremely opportunistic. While normally associated with the sea, they can be seen following ploughs to feed on earthworms, on rubbish dumps and pirating snacks at seaside resorts: in 2016 a Herring Gull's enthusiasm for free food ended in it falling into a vat of tika masala at a UK food factory, emerging unscathed, but a delightful shade of orange and smelling interestingly. Larger gulls are also efficient predators, with the larger species feeding on smaller birds and chicks. Some gulls will cannibalise chicks in their own colonies: studies have suggested that in Herring Gull colonies up to 25% of the chicks will be consumed by adult birds each year. Most extraordinary of all, Ivory Gulls feed on the faeces of Polar Bears, especially in winter, though they also take crustaceans at all times.

Gulls are voracious feeders. They have large crops that aid the feeding of chicks by regurgitation. In some species adults have a red gonys spot on the lower mandible – the chicks peck at this to stimulate regurgitation. Aiding their success is superb flying ability. They are excellent at soaring and gliding, and have the high manoeuvrability essential for the cliff nesting that most favour (though some are ground nesters and, to the surprise of first-time viewers, Bonaparte's Gull nests in trees.

Gulls are chiefly birds of colder waters, and are particularly numerous in the northern hemisphere. The majority are primarily marine, though coastal during the breeding season of course. In winter they are no longer tied to land and some species become pelagic. This section begins with the trio of much-sought after Arctic breeders, Ross's, Ivory and Sabine's Gull. →

Ivory Gull   *Pagophila eburnea*
Status: IUCN: Near threatened (population believed to be declining, though difficulty of accurate counts means this may not be the case).
One of the Arctic specialities that all visitors will want to see. Often viewed above sea ice, occasionally appearing ghost-like from out of the mist, a magical, ethereal sight. A two-year gull (see a later footnote for details of gull plumage). The only pure white gull (all other white gulls will be albinos). Adults are white, but the face is sometimes red after feeding on carrion. The bill is blue-grey with a yellow, orange or red tip. The legs and feet are black. Summer and winter plumages are similar.

Ivory Gulls breed in north and east Greenland, on Jan Mayen, Svalbard, and Russia's Arctic islands (apart from the southern island of Novaya Zemlya). Breeds on all Canada's northern Arctic islands. Moves south with the advancing ice edge.

Ross's Gull   *Rhodostethia rosea*
If the Ivory Gull is the most magical of Arctic gulls, Ross's must qualify as the most beautiful. The two are equally elusive, much more so than Sabine's Gull, the third of the Arctic triumvirate. A two-year gull. Breeding adults have white heads with, uniquely, a thin black necklace extending from throat to mid–nape (and wider towards nape). The small, thin bill is black. The underparts are white, usually with a rosy-pink tinge, strongest on the breast and belly. The upperparts are pale blue-grey, with a broad white trailing edge. The legs and feet are red. In winter adults lose the rosy-pink tinge, and the black necklace reduces to a black streak on the sides of the neck.

Ross's Gull's breeding range is poorly studied. Known to breed at sites on the southern Taimyr Peninsula, on the Lena and Kolyma deltas and in Chukotka. Has bred in north-east Greenland and on Svalbard but has not become established in either location. Remarkable the gull has also bred at Churchill on the southern shore of Hudson Bay, but breeding there has been sporadic. It is possible Ross's Gull breeds at remote sites on Canada's High Arctic islands. In winter the birds are seen in the Bering Sea. Flocks were once regularly seen passing Barrow during the autumn, but they are very much less common now.

a *Ivory Gulls feeding on a Polar Bear kill.*
b-d *Ivory Gulls.*
e *Nesting Ross's Gull being attacked by an Arctic Tern.*
f-h *Ross's Gulls.*

Sabine's Gull *Xema sabini*

Two-year gull. Breeding adults have dark grey hoods extending to the upper nape. On the nape the hood has a thin black rim. The short bill is black with a yellow tip. The underparts are white, occasionally tinged pink. The rump and uppertail are white. The tail is white, with a shallow fork. The upperparts are blue-grey. The upperwing is tricoloured, blue-grey, white and black, each colour forming a triangle which can cause confusion with juvenile Black-legged Kittiwakes, but the tail of the latter has a terminal black band. The Sabine underwing is white, occasionally with pale grey triangles and black tips on the primaries. The legs and feet are dark grey. In winter adult loses the hood but the nape is extensively smudged black.

Sabine's Gulls breed in northern Greenland, on Svalbard, in northern Chukotka and on Wrangel Island. Also in western and northern Alaska, northern Yukon, Banks, Victoria and Southampton Islands, and on western Baffin Island. In winter the birds migrate – the only truly migratory Arctic gull – and are pelagic, feeding in the Benguela Current off south-western Africa, and the Humboldt Current off north-western South America.

Iceland Gull *Larus glaucoides*

Four-year gull. Breeding adults have white heads. The bill is yellow (often tinged green) with a red gonys spot. The underparts are white, as are the rump and tail. The upper and lower wings are pale grey or blue-grey, with a distinct white trailing edge. The legs and feet are pink. Winter adults are as breeding but with some dark streaking on the head (though this is occasionally absent). Kumlien's Gull (often considered a sub-species, but regarded as a full species by some authorities – see below for further details on speciation) has darker grey wing-tips.

Despite the name, the bird is only seen in Iceland during the winter. Breeds in west and south Greenland and, more rarely, on the south-east coast. Breeds on northern Baffin Island and south-west Ellesmere. Kumlien's Gull breeds on southern Baffin Island, western Southampton Island and the extreme northern tip of Quebec. In winter the birds are seen across the North Atlantic from Newfoundland to Iceland, but rarely further east. Wintering birds are both coastal and pelagic.

Thayer's Gull *Larus thayeri*

Four-year gull. Breeding adult patterned as Iceland Gull, but in general darker grey on the upperparts (usually the same colour, or a little darker, than Herring Gulls). The wing-tips show black areas that are absent on Iceland Gulls, the overall effect being of a streaked black-and-white wing-tip (*smithsonianus* Herring Gulls have solid black wing-tips, apart from the white tips of the last two primaries). The bill is yellow with a red gonys spot. The legs and feet are pink. In winter adults are as breeding, but the head, neck and upper breast are lightly streaked grey-brown.

Thayer's Gulls breed on the western shore of Hudson Bay and the Canadian Arctic islands from Banks east to northern Baffin and north to Ellesmere Island, though apparently absent from the Parry Islands.

Biogeographers argue about the status of Iceland and Thayer's gulls. It is assumed that an ancestral form was isolated by an ice sheet which separated eastern and western populations which evolved into Iceland and Thayer's gulls respectively. However, some authorities consider the two taxa as subspecies. An extra complication is afforded by Kumlien's Gull (*Larus kumlieni*), considered a full species by some and a subspecies of Iceland Gull by others. Add to this mix of species and subspecies the occasional hybrids and the result is a recipe for argument and discussions that will continue for some time. As noted overleaf, there is also controversy over the status of the northern Herring Gulls and Common (Mew) Gulls.

a–c, e Adult Sabine's Gull, showing the distinctive underwing (b) and upperwing (e) colour patterns.
d Adult Thayer's Gull.
f, g Iceland Gulls. Photo 'f' shows the comparative size to a Glaucous Gull (left).

Herring Gull

Four-year gull. Breeding adults have white heads and yellow bills with a red gonys spot. The underparts and tail are white, the upperparts blue-grey. Wings have white trailing edges and black-and-white tips. The legs and feet are pink. Winter adults are as breeding, but the head and neck streaked grey-brown.

Larus argentatus (Photos 163a and b) breeds on Iceland and in northern Fennoscandia: several subspecies are considered to exist. *L. vegae* breeds across Russia: Several subspecies are considered to exist. *L. smithsonianus* (Photos 163c and d) breeds across North America, including Southampton Island and southern Baffin Island. No subspecies have been identified. Some authorities consider the three species, virtually identical apart from subtle differences in plumage and eye/eye ring colour, to be conspecific, subspecies of *L. argentatus*. In all cases birds move south in winter to wherever food can be obtained.

Common Gull/Mew Gull

Three-year gull. Breeding adults have white heads and a yellow bills, usually with a brighter tip and greenish tinge to base. The underparts and tail are white. The upperparts and upperwing are grey. There is a white trailing edge and the outer primaries are black with white 'mirrors'. The legs and feet are yellow or greenish-yellow. In winter adults are as breeding, but the head and neck are streaked grey-brown

Larus canus (Photo 163f) breeds on Iceland (in limited numbers: the species first bred there in 1936 and the range is expanding), in northern Fennoscandia and across Russia (though Russian birds may be a subspecies), but always subarctic. *L. kamtschatschensis* breeds in southern Chukotka and Kamchatka. *L. brachyrhynchus* (Photo 163e) breeds throughout Alaska (but is rare in the north), in Yukon and North-West Territories (again, rare in the north). As with Herring Gulls, the differences between the various forms are subtle, and some authorities consider all the birds to be conspecific. However, mtDNA analysis suggests that North American gulls (Mew Gulls) differ sufficiently to warrant full species status.

Lesser Black-backed Gull *Larus fuscus*

Four-year gull. Similar to Herring Gull, but a more slender bird with a dark grey or black upperwing. The wing has broken white leading and trailing edges. The legs and feet are yellow. In winter adults are as breeding, but the head and neck are streaked grey-brown and the legs and feet are duller (Photos 163g and h).

Breeds in southern Greenland, on Iceland, in northern Fennoscandia and Russia east to the Taimyr Peninsula. Increasing numbers are now seen in North America, but it is not yet an established breeding species. In winter Greenlandic and Icelandic birds move to Iberia and north-west Africa, while Scandinavian and European Russian birds head for the eastern Mediterranean, the Arabian Sea and east Africa.

→The speciation of gulls is a fertile subject for biogeographers. Herring Gulls and Lesser Black-backed Gulls are sympatric where they overlap, but they do occasionally interbreed successfully. It is suggested that these species arose when two populations of an ancestral gull were separated during one of the Pleistocene ice ages. The ancestral form is thought to have been yellow-legged and confined to central Asia. It evolved into the Lesser Black-backed Gull and spread west towards Europe. The second population was isolated in north-eastern Asia. It evolved pink legs and spread east, crossing North America. However, this simple story is complicated by the curious distribution of the species in the Palearctic. In addition to breeding in north-east Asian Russia, Herring Gulls also breed around the White Sea and as far east as the Taimyr Peninsula, as well as throughout northern Europe. Herring Gulls also became established in Iceland in the 1920s, just as the Lesser Black-backed Gull did, and have occasionally bred in Greenland (arriving from both Canada and the eastern Atlantic). Speciation of the two gulls is further enlivened by the existence of subspecies of each, these being classified as full species by some authorities. There are also several Lesser Black-backed Gull subspecies, one of which, a larger bird with paler wing-tips (but with black wing-tips) that breeds from the White Sea to the Taimyr Peninsula, is considered by many a full species, Heuglin's Gull *L. heuglini*. To add yet more complication, some authorities consider that not only is Heuglin's Gull a full species, but that it also has subspecies!

Black-legged Kittiwake *Rissa tridactyla*

Three-year gull. Breeding adults haves white heads. The bill is yellow. The underparts, rump and notched tail are white. The upperparts are grey. The upperwing is tricoloured, the inner wing being grey (as the mantle) with a white leading edge, the primaries pale grey and the wing-tip black. The legs and feet are black. In winter, adults are as breeding, but with a dark grey crescent from the ear-covert and, often, pale grey on the nape, and the bill is duller, often green-yellow. Juveniles have distinctive black barring on the wings in the form of an M (or W, depending on the direction of flight!). The bird's curious name derives from its call, a trisyllabic *kitt-ee-wake*.

Black-legged Kittiwakes breed in west, south, east and north-east Greenland, on Iceland, Jan Mayen and Svalbard, in northern Fennoscandia and at a limited number of sites on Russia's northern mainland (though common in Chukotka and Kamchatka). Also breeds on all Russia's Arctic islands. Breeds on all Bering Sea islands, western Alaska (but rare on the north coast) and eastern Canada, including eastern Baffin Island. Pelagic, moving ahead of the sea ice in winter to feed in the North Atlantic and North Pacific.

Glaucous Gull *Larus hyperboreus*

Four-year gull. Breeding adults have white heads. The bill is yellow with a red gonys spot. The underparts are white. The upperwing is pale grey with a white trailing edge. The rest of the upperparts are white. In winter, adults are as breeding, but the head, neck and upper breast are streaked grey-brown.

Glaucous Gull is a true Arctic breeder and a formidable predator throughout the region. Breeds on Greenland, Iceland, Jan Mayen and Svalbard, in northern Fennoscandia and across Russia, including all Arctic islands. Breeds across the northern mainland of North America and on all Canadian Arctic islands to northern Ellesmere. In winter the birds move south, but only as required by the sea ice. Wintering birds are both coastal and pelagic.

Greater Black-backed Gull *Larus marinus*

Four-year gull. The largest gull. Breeding adults have white heads. The bill is yellow with a red gonys spot. The underparts are white. The back and upperwing are dark grey, with a white trailing edge and black primaries. The outer primaries have white 'mirrors'. The underwing is banded white, pale grey and darker grey, with a white trailing edge. The legs and feet are pale pink. In winter, adults are as breeding, but the head and neck are streaked grey-brown. This huge gull often kill auks and other small seabirds in flight, stabbing them with its bill.

Greater Black-backed Gulls breed on Canada's Labrador coast, on the central west coast of Greenland, on Iceland, Jan Mayen and Svalbard (where it became established only in the 1930s), in northern Fennoscandia, on the southern island of Novaya Zemlya, and on the nearby mainland coast of Russia. Partially migratory, but rarely seen beyond the limit of the continental shelf.

→Gulls are long-lived and take time to mature, the smaller gulls sometimes breeding in their second year, the larger ones not until they are four or five years old. Juveniles do not acquire their adult plumage for several years, and as the differences in juvenile plumages can be slight, identification of juvenile forms can be tricky. No attempt has been made to identify the changes in juvenile plumages in the descriptions here: whole books have been devoted to these changes. However, the time taken by juveniles to attain adult plumage is given. The species are classified as two-, three- or four-year gulls, the time being the number of winters before a juvenile acquires adult plumage, i.e. a four-year gull shows its adult plumage in the fourth winter of its life and will show full adult breeding plumage in its fourth spring. The sexes are essentially similar in all species.

a-c Black-legged Kittiwakes.
d Glaucous Gull.
e Black-backed Gull.

Slaty-backed Gull *Larus schistisagus*

The only dark-backed gull in the Bering Sea. Four-year gull. Breeding adults have white heads. The bill is yellow with a red gonys spot. The underparts, rump and tail are white. The upperwing is slate-grey with a broad white trailing edge. The primaries are dark grey at the base, darkening to black towards the tip, the outer primaries having white crescents that form an often distinct pattern known as the 'string of pearls'. When distinct this 'string' is both diagnostic and delightful. Winter adults are as breeding, but with some grey-brown streaking to the crown, and heavy streaking to the nape and sides of the neck.

Slaty-backed Gulls breed in far eastern Chukotka, on Kamchatka and along the northern coast of Sea of Okhotsk. Occasionally seen in Alaska (e.g. Pribilof Islands), but breeding not proven and certainly not established. In winter the gulls are seen off coasts as far south as Japan.

Bonaparte's Gull *Larus philadelphia*

Named after the French zoologist Charles Lucien Bonaparte, a nephew of Napoleon Bonaparte. Two-year gull. Breeding adults have a black hood extending to the rear crown. The bill is black, the legs dark pink. The underparts are white, as is the tail. The back and inner wing are pale blue-grey, the outer wing white with a black tip. There are also black tips to the outer primaries, creating a black trailing edge. In winter adults lose the hood, though black smudging often remains, and the legs and feet are a dull pale pink.

Bonaparte's Gulls breed in southern Alaska, but it is rare on the western and northern coasts. Also breeds in northern Yukon, then southerly to the southern shores of Hudson and James Bays. In winter the birds move the east and west coasts of the United States and to the Great Lakes.

Glaucous-winged Gull *Larus glaucescens*

Four-year gull. Breeding adults have white heads. The bill is yellow with a red-orange gonys spot. The underparts are white. The legs and feet are pink, sometimes pinkish-purple. The back and mantle are mid-grey, with a blue tinge. The rump and tail are white. The upperwing is mid-grey (again blue-tinged) with a white trailing edge. The outer primaries have mid-grey ends with white tips. In winter, adults are as breeding, but the head and neck are streaked grey-brown.

Glaucous-winged Gulls are mainly non-Arctic, but breed in southern and western Alaska, in southern Chukotka, on Kamchatka and the Commander Islands. In winter the birds are resident or migratory, some moving as far south as Baja California and Japan.

Red-legged Kittiwake *Rissa brevirostris*

Status: IUCN: Vulnerable (a significant population decline in the 19702-1990s for a species which nests only at a few locations. The decline on the Pribilofs now seems to have halted, the population being stable.)

Three-year gull. Breeding adults are similar to Black-legged Kittiwakes, but the grey upperparts are darker, with less contrast between the pale inner wing and the darker outer wing. The black wing-tip is broader and less conspicuous against the rest of the dark wing. The white band on the trailing edge of the wing is also broader. The legs and feet are bright red. Winter adults have an ear crescent as in Black-legged Kittiwake, but it is much darker.

Red-legged Kittiwakes have a highly restricted breeding range, breeding only on the Pribilof Islands, Buldir and Bogoslof islands in the Aleutians, and on the Commander Islands. It is to be hoped that the volcanic eruptions on Bogoslof in late December 2016, which have split the island into two, will not reduce the gulls breeding options. Essentially pelagic, with wintering birds moving to the North Pacific.

a Slaty-backed Gulls, one illustrating the 'string of pearls' created by the mirrors on the primaries.
b Adult Glaucous-winged Gull.
c Adult Slaty-winged Gull.
d Adult Bonaparte's Gull.
e The extraordinary sight of a gull nesting in a tree. Bonaparte's Gull breeding at the treeline on the southern shore of Hudson Bay.
f Red-legged Kittiwake nesting among guillemots, Pribilofs.

Terns

Terns are found on all the continents, though they are mostly tropical birds. The three Arctic breeders belong to the Sterna group of black-capped terns, slender birds with tapering wings and deeply forked tails. They are fish and marine invertebrate eaters, hunting by plunge diving to take prey just below the surface as the birds do not swim underwater. Terns often seek shoals that have been driven close to the surface by predatory fish or aquatic mammals. Arctic Terns also make piratical attacks on conspecifics, other terns and auks.

Terns are monogamous, the pair bond life-long, or long-lived. They are gregarious at all times, with colonial breeding leading to nests so close together that a pair must defend its small territory. Arctic Terns are extremely aggressive to intruders and will attack, and strike, anything, including humans, who roam too close. At the nest two to three eggs are laid, these being incubated by both birds. Chicks are downy and semi-precocial. They are fed by both birds. Terns are long-lived and may be five years old before first breeding.

Arctic Tern *Sterna paradisaea*
Breeding adults have black caps (to the eye) and black napes. The rest of the head is white. The bill is bright red. The underparts are pale grey or blue-grey. The rump and tail are white. The upperbody is blue-grey. The legs and feet are red (Photos 169a-c). In winter adult have a smaller, ragged black cap, and the legs and feet are darker, sometimes almost black. Non-breeding would be a better term than 'winter' as Arctic Terns migrate to Antarctica after breeding and so experience two summers each year. It was always assumed that these flights might involve journeys of around 17,500km in each direction. It is now known that this assumption seriously underestimated the true scale of the trips. Satellite tagged birds have been found to fly along the west coast of Africa, and then to turn into the Indian Ocean, some even reaching Australia and flying along the south coast to Melbourne before turning south. In 2016 a bird tagged on the UK's Farne Islands clocked a total round trip flight of 96,000km. With a probably lifetime of around 30 years, a tern might therefore travel over 3 million kms (about the same as Earth to the Moon and back four times). With limited ability to soar and glide, the birds must flap most of that distance. They can, of course, feed along the way, but some Russian birds migrate across continental land masses, with limited opportunities to refuel.

Breeds in Greenland, Iceland, Jan Mayen, Svalbard, Fennoscandia and across Russia to the Bering Sea, including Russia's Arctic islands. Breeds across North America from Alaska to Labrador, including the Canadian Arctic islands.

Aleutian Tern *Sterna aleutica*
Breeding adults have black caps to the eye and a black nape, but a white forehead (Photo 169d of an incubating female). The rest of the head is white. The bill is black. The throat is white, the rest of the underparts mid-grey. The rump and deeply forked tail are white. The upperparts are mid-grey. The legs and feet are black. In winter the birds lose the black crown.

Breeds in southern Chukotka, on Kamchatka and the Commander islands, in western Alaska and on the Aleutians, but are uncommon throughout the range. Resident, but pelagic outside the breeding season.

Common Tern *Sterna hirundo*
Breeding adults are as Arctic Tern, but the bill is black-tipped, the wing shows a broad, but diffuse, grey band towards the tip but lacks the black trailing edge, and has longer legs (Photo 169f, the Common Tern is on the left: the other birds are Arctic Terns). The latter are noticeable when the birds are perched, the Arctic Terns usually looking as though they are resting on their bellies. In winter the black cap becomes duller, and grey-brown towards the bill: the bill becomes black with a red base.

Essentially non-Arctic, but breeds at a small number of sites in north Norway, on the southern shore of the White Sea, and on Kamchatka. In winter the birds head south to the coasts of southern Africa, India, and waters around Indonesia. The subspecies on Kamchatka, *S. h. longipennis*, differs in having a black bill and dark red or black legs (Photo 169e). The differences are such that the bird was considered a full species in the past.

Auks

Auks are often referred to as the northern equivalent of the penguins, though the anatomical adaptations of auks are much closer to those of the diving-petrels of the Southern Ocean. However, the auks do share some penguin-like characteristics, adaptations that have resulted from the two bird families having faced similar evolutionary pressures. The legs are set well back on the body and the feet are webbed to act as efficient paddles for swimming. The larger auks adopt the same upright stance as the penguins, though some of the smaller auks lie on their bellies on land. The wings are short and used for underwater propulsion, but while penguins have lost the power of flight, auks have had to retain this ability as northern regions have terrestrial predators. But short wings mean high wing loadings, so auks fly with furiously beating wings. The two families share an extraordinary diving ability. Only the larger penguins dive deeper than the larger guillemots: one confirmed dive of a Brünnich's Guillemot reached 210m. Most dives are to shallower depths, but the larger auks regularly dive to 60m and can stay submerged for up to three minutes. Auks can also swallow food underwater, an adaptation that allows a longer dive time.

Auks are long-lived birds, their chicks taking time to mature and not breeding until they are two or three years old; this applies even for the smaller auks.

Though the auks are a northern group, the distribution of species is highly asymmetric, with many more in the Pacific than in the Atlantic. The reasons for this are still debated, but it is likely that in part it derives from auks having evolved in the Pacific: the only Atlantic species without a Pacific equivalent is the Razorbill (though there is also no Pacific equivalent of the now-extinct, flightless Great Auk). Auks are birds of the continental shelf, only the puffins being found in deeper waters, particularly in winter, when they are truly pelagic. →

Razorbill *Alca torda*

Status: IUCN: Near threatened (population decline, particularly on Iceland, apparently linked to sandeel stock crashes associated with sea temperature rise resulting from climate change. Severity of winter storms also now causing high mortality).
Breeding adults have black heads with a thin white stripe from bill to eye. The large (and razor-sharp) bill is black with a vertical, curved, white stripe crossing the mandibles in front of the nostril. The upper breast is black, the rest of the underparts white. The upperparts are black, apart from a white trailing edge on the secondaries. In winter adults are as breeding, but the throat, face and sides of the neck are white, occasionally smudged black, and the white facial line is lost.

Razorbills breed on south-west Baffin Island, in northern Quebec and Labrador, west-central and south-west Greenland, on Iceland, Jan Mayen, Bear Island and (in small numbers) Svalbard. Breeds in northern Fennoscandia and there are colonies on the western White Sea coast, but the bird is absent from the rest of the Russian Arctic. Partially migratory, moving to waters off north-eastern United States and Canada, to the western Atlantic, the North Sea and western Mediterranean.

Little Auk (Dovekie) *Alle alle*

Related to the guillemots and Razorbill despite its similarity to the smaller auks. Breeding adults are patterned as the larger auks, with very dark brownish/black upperparts. There are white streaks on the upperbody. The short bill is black. In winter, the head and upper breast are white, as in the guillemots.

Little Auks breed in west-central and northern Greenland, on Iceland, Jan Mayen, Svalbard, Franz Josef Land, Novaya Zemlya and Severnaya Zemlya. There is also a small colony on western Baffin Island. Remarkably, in recent years small but expanding colonies have been found on the Diomede Islands and St Lawrence Island in the northern Bering Sea. Partially migratory, moving south to cold currents of the North Atlantic and North Pacific.

a–d Razorbills.
e–g Little Auks.

Common Guillemot (Common Murre) *Uria aalge*

Breeding adults are patterned as Razorbills, but with very dark brown head and upperparts, the flanks showing dark brown smudging. The dagger-like bill is black. The wing has a thin white trailing edge. In the two guillemots and the Razorbill there is a narrow channel in the feathers leading away from the eye which seems to aid the flow of water past the eye when the birds swim rapidly underwater. In some North Atlantic colonies this channel is picked out in white: these birds also have a white orbital ring. These birds are known as 'bridled' guillemots. The percentage of bridled birds increases as the observer travels north. There are no bridled birds in the south of the species' range (e.g. Iberia), but around 50% of birds are bridled in Iceland, Svalbard and Novaya Zemlya. In winter the cheeks, throat and upper breasts are white. Guillemots are long-lived birds: in 2016 one trapped on Scotland's Isle of Canna was found to have been ringed in 1978, making it 38 years old. Several others were found to be 34-36 years old.

Breeds circumpolar, but with a discontinuous distribution, being seen in south-west Greenland, Iceland, Jan Mayen, Svalbard and the southern island of Novaya Zemlya. Breeds in northern Fennoscandia, but absent from the Russian coast except for eastern Chukotka and Kamchatka. Breeds on the Aleutians and islands of the Bering Sea and on the west coast of Alaska, but not on the northern North American mainland (apart from northern Labrador) or the Canadian Arctic islands. Dispersive rather than migratory, with some birds being resident and others moving to nearby, but more southerly waters.

Brünnich's Guillemot (Thick-billed Murre) *Uria lomvia*

Breeding adult patterned as Common Guillemot, but the upperparts are much darker, essentially black, though paler on the head. In mixed colonies the difference in colour of the upperparts is striking. No 'bridled' form has been observed. The flanks are unsmudged. The bill is shorter than Common, with a downcurved culmen and a pale streak along the basal edge of the upper mandible. In winter there is less white on the head than in Common Guillemots.

Brünnich's Guillemot tend to breed further north than their Common cousins, being seen on west and north Greenland, Iceland, Jan Mayen, Bear Island and Svalbard, in northern Fennoscandia, on all Russia's Arctic islands. There are also isolated colonies on the northern Russian mainland and on eastern Chukotka and Kamchatka. Breeds on the Aleutians and Bering Sea islands, in west and north-west Alaska, on Baffin and Ellesmere islands, and in northern Quebec and Labrador. Dispersive in winter, to the North Atlantic and North Pacific.

→ Auks show marked differences in bill shape, these related to the choice of prey. Fish-eaters tend to have dagger-like bills as for other piscivorous birds, while plankton-feeders have shorter, wider bills. The Parakeet Auklet, which feeds on jellyfish (as well as crustaceans), has a curious, rounded bill, rather like a scoop. The bills of the Razorbill and the puffins are laterally compressed and seem, particularly for the puffins, to be important in mating displays, the male often showing it to rivals, laying it across his white breast in territorial disputes. Puffin bills are encased in nine plates that are shed during the autumnal moult, the 'new' bill being much more subdued. Puffins also shed their flight feathers simultaneously, something they share with the other larger auks. This means that the birds are flightless for a period, but the wing loadings of these birds mean that the successive lost of flight feathers would leave them flightless, and therefore more vulnerable to predation, for a longer period.

Some auks have distinctive head plumage that acts in a similar way to the puffin's bill, the plumes being displayed to rivals or potential mates. The Tufted Puffin has both crests and a splendid bill. Apart from the bills and head plumage, auks have other mating displays; these usually involving head-shaking and bowing and, after the pair has formed, bill-nibbling or clacking as bond reinforcement. The murrelets, which find walking difficult, tend to have sea-based displays, often involving parallel swimming. The pair bond is monogamous and long-lived or life-long.→

a Common Guillemots.
b Common Guillemots showing bridled and 'standard' forms.
c-e Brünnich's Guillemots.

Black Guillemot *Cepphus grylle*

Breeding adults are black overall apart from a white patch on the upperwing. The thin, dagger-like bill is black. The legs and feet (and the mouth) are bright red. In winter adults are mottled black-and-white, a complete contrast to the summer plumage.

Black Guillemots breed on west and north Greenland, on Iceland, Jan Mayen, Svalbard, Franz Josef Land, Novaya Zemlya and Severnaya Zemlya, northern Fennoscandia, and the Taimyr Peninsula. Also breeds on the New Siberian Islands and Wrangel Island, and the adjacent Russian coast to eastern Chukotka. Breeds in north-west and north Alaska, on Canada's northern coast from the Yukon to Franklin Bay, on northern Hudson Bay coasts and adjacent islands, western Baffin Island and southern Ellesmere Island. Resident or dispersive, staying as far north as conditions allow.

Pigeon Guillemot *Cepphus columba*

Breeding adults are as Black Guillemots but ranges do not overlap. However, the Pigeon's white wing-patch is partially crossed by a black wedge. The thin, dagger-like bill is black. The legs and feet (and mouth) are bright red. In winter adults are as Black Guillemots.

Pigeon Guillemots breed on eastern Chukotka, eastern Kamchatka and the Kuril and Commander Islands, on the Aleutian islands and in west, south and south-east Alaska. Resident, but moves ahead of the sea ice.

Spectacled Guillemot *Cepphus carbo*

Poorly studied and essentially non-Arctic, but may be seen at the Arctic 'boundary'. Breeding adults are as Black and Pigeon Guillemots, but larger, dark brown rather than black, and with a prominent white eye patch and two smaller white patches above and below the bill base. There is no white wing-patch. The bill is black. The legs and feet are red. In winter adults are as Black and Pigeon Guillemots.

Spectacled Guillemots breed on coasts of the Sea of Okhotsk and north-west Kamchatka. Resident or with a limited winter movement south.

→ Most auks are colonial breeders, though some colonies are rather loose, and several of the smaller Pacific auks are solitary nesters. Nests vary considerably within the family: some auks nest on cliffs, laying eggs directly onto a ledge with little or no nesting material. Some nest in burrows, usually excavating these themselves. Kittlitz's Murrelet occasionally breeds far inland, a curiosity for a bird that finds walking so troublesome: it makes a scrape in bare tundra, sometimes at altitudes to 600m and often in snowfields. The Marbled Murrelet is exceptional in nesting in trees: its nest sites are extraordinary, the egg being placed on an old, wide branch or where epiphytes create platforms. Even more remarkably, the birds are usually nocturnal at the nest sites, flying through the woodland in darkness.

In most species a single egg is laid (two is more common in a few species). The eggs of the larger guillemots are pyriform i.e. pointed at one end, an adaptation to minimise the chances of the egg rolling off the narrow nesting ledge. The eggs are of highly variable colour, particularly for the Common Guillemot, which has the most variable egg of any bird: the base colour varies from white to a beautiful turquoise, with a patterning of dark scribbles and splotches. The egg or eggs are incubated by both birds.→

a-c Black Guillemots.
d   Spectacled Guillemot.
e, f Pigeon Guillemots.

Kittlitz's Murrelet *Brachyramphus brevirostris*

Status: IUCN: Near threatened (paucity of surveys means population not well documented, but thought to be in moderate decline from low numbers).

A rare, small auk with legs set far back so that standing and walking is difficult: the bird rests on its belly on land. Breeding adults have mottled grey and brown heads and upperbodies, the underparts mottled grey and golden-brown apart from white undertail-coverts. The black bill is short and feather-covered above the nostril (Photo 177c). In winter adults are grey and white, but retain dark wings.

Kittlitz's Murrelets breed on Wrangel Island, the coasts of eastern Chukotka and north-western Kamchatka, on the Aleutian islands, and in isolated colonies on the west and south Alaskan coasts. The 1989 *Exxon Valdez* spillage is thought to have killed 10% of the world population. Northern birds fly south in winter to join resident southern birds.

Marbled Murrelet *Brachyramphus marmoratus*

Status: IUCN: Endangered (logging of old-growth forests restricts nest sites).

Breeding adult almost identical to Kittlitz's Murrelet, but darker overall and with darker undertail-coverts. The bill is also longer. In winter adults have grey crown and nape, grey and white upperparts and white underparts (Photo 177a).

Nests in trees. This very particular nesting requirement is best satisfied by old-growth forest, so the birds are vulnerable as much of this has been logged and logging still continues. An Arctic fringe species, Marbled Murrelets breed in south-western Alaska, on the Alaskan Peninsula, and on the eastern Aleutians. Resident.

Long-billed Murrelet *Brachyramphus perdix*

Status: IUCN: Near threatened (as with Marbled Murrelet, logging of old-growth forests restricts nest sites).

Formerly considered a subspecies of the Marbled Murrelet, but now given full species status. Plumage very similar to the Marbled Murrelet, but the throat is paler and the overall colour greyer. Seen together, the longer bill is diagnostic (Photo 177b). Easier to distinguish in winter as the grey of the nape extends to the rear cheeks.

Long-billed Murrelets breed on Kamchatka, the coasts of the Sea of Okhotsk and on Sakhalin. In winter the birds move to waters off southern Sakhalin and northern Japan.

Ancient Murrelet *Synthliboramphus antiquus*

Delightful little birds, their name deriving from the white streaks on the black head that give the look of a grey-haired or balding old man. The throat is black, the rest of the underparts white, heavily streaked grey on the flanks. The underwing is white with slate-grey flight feathers. The upperparts are dark slate-grey (Photo 177d). In winter, adults lose most of the white head-streaks, but the amount of black on the head reduces.

Ancient Murrelets are nocturnal, as are the Marbled and Long-billed Murrelets, and so are most likely to be seen when leaving or returning to their nests sites, though they may be encountered at sea.

Ancient Murrelets breed on islands off the south coast of the Alaskan Peninsula, on the Aleutian and Commander Islands, on north-west and south-east Kamchatka, on the north coast of the Sea of Okhotsk, and on Sakhalin and the Kuril Islands. In winter the birds are seen in southern Alaskan waters, and off south-east China, Japan and Korea.

Crested Auklet *Aethia cristatella*

Breeding adults have very dark grey heads and necks, with a forward-curved crest of the same colour and white plumes drooping from behind the eye. The bill is bright orange and has basal tubercles of the same colour. The underparts are mid-grey. The upperparts and tail are dark slate-grey (Photos 177e and f). In winter, adults lose the white eye plumes and the bill-base tubercles. The bill is also duller.

Crested Auklets breed on the Aleutians and Bering Sea islands, on Chukotka near Providenya, on both coasts of northern Kamchatka, on the northern coast of the Sea of Okhotsk, and on the Commander Islands and Sakhalin. In winter the birds are seen in a band stretching across the North Pacific from northern Japan to the southern Alaskan Peninsula.

Parakeet Auklet *Aethia psittacula*
Breeding adults have a dark greyish-brown head with a long white plume curving down from behind the eye towards the nape. The bill is bright reddish orange. The breast is dark grey, the rest of the underparts white. The upperparts and wings are dark grey (Photos 179a-c). In winter adults have duller bills but retain the head plumes.

Breeds on islands off southern Alaska, on Aleutian, Bering Sea and Commander islands, on eastern and southern Chukotka, islands of the Sea of Okhotsk and the Kuril Islands. Some southern birds are resident, northern birds move south to join others in a broad band from northern Japan to the west coast of the United States.

Least Auklet *Aethia pusilla*
Breeding adults have dark and pale morphs, and intermediate forms. In all forms the upperparts are the same, the head dark grey (or grey-brown) with short white forehead plumes and longer white plumes behind the eye. The bill may be black with a red tip or red with a black base. A black knob on the culmen is very prominent in some birds but virtually invisible on others (Photos 179d). In winter the underparts are white and plumes are lost.

Breeds on islands off the southern Alaskan Peninsula, on the Aleutian, Bering Sea and Commander islands, on Chukotka near Providenya, on islands near Kamchatka and islands in the Sea of Okhotsk. Resident, but moves south ahead of the sea ice.

Whiskered Auklet *Aethia pygmaea*
The head plumes of breeding adults are the most ornamented of the auklets, with forward-curving dark grey forehead plumes, white plumes from behind the eye towards the nape, and two plumes from the bill base that curve above and below the eye (Photo 179f). In winter adults are as breeding, but with vestigial plumes.

The rarest of the northern auks, but Arctic-fringe only, breeding on islands in the northern Sea of Okhotsk, on the Kuril and Commander islands, and a few Aleutian islands. Essentially resident in winter, but move further offshore in winter.

Cassin's Auklet *Ptychoramphus aleuticus*
Status: IUCN: Near threatened (vulnerable to effects of climate change, but also to terrestrial and avian predators, numbers of which have increased near nest sites).

Photo 179e. Arctic-fringe species. Breeds on some Aleutian islands and islands south of the Alaska Peninsula. In winter seen off south coasts of Aleutian islands and western USA.

Rhinoceros Auklet *Cerorhinca monocerata*
Breeding adults similar to Parakeet Auklet, but larger, and with a vertical horn (up to 25mm tall: hence the name) arising from the lower mandible of stubby orange bill (Photo 179g).

Arctic-fringe species breeding on south coast of Alaska Peninsula and nearby islands, and limited islands of Sea of Okhotsk. Winters on southern US west coast, Japan and Korea.

→Development strategies of auk chicks are as varied as those for nesting. The chicks of most species are semi-precocial and are fed until they are at, or close to, adult weight, at which point they become independent. Little Auks differ, with fledglings joining the adult male on the sea for a short time. In guillemots and Razorbills chicks are fed until they are about 25–35% adult weight. The still-flightless chick is then encouraged to leave the nest ledge by adult calls, and glides to join its parents at sea. It is assumed this arises because the adults, which have high wing loadings, cannot carry the quantities of food required to allow continued development at the nest. The disadvantage is the need for the chick to reach safety at sea by gliding while skuas and gulls seek to snatch them from the sky: if it cannot gain the sea, the chick may either not survive the crash landing on the ground or the hurried scramble to the sea as the base of nesting cliffs are often patrolled by Arctic Foxes. A third strategy has evolved among some smaller auks (including Ancient Murrelets). Here the chicks are precocial and can swim within two or three days, so the join adults at sea where they are fed for several weeks before being abandoned. This strategy has presumably developed to save the adults the energetic cost of flying to the nest, though together with nocturnality, which is a feature of some smaller auks, it may also be a defence against avian predators.

(Atlantic) Puffin *Fratercula arctica*
Status: IUCN: Vulnerable (highly susceptible to climate change as sea temperature rises move major food source away from potential nesting sites).
The three puffins are unmistakable – the clown princes of the bird world.

Breeding adult Atlantic Puffins have black crowns, napes and collars, the rest of the face being white or pale grey. The forward half of the large bill is red, the basal half blue-grey, these colours outlined and separated in yellow. The underparts are white. The upperparts and tail are glossy black. The upperwing is black, the underwing dark grey and silver-grey. The legs and feet are reddish orange. In winter adults lose the bill plates, the distal section becoming dull red, the basal section dull brown. The facial disc becomes grey, and much darker in front of the eye.

Atlantic Puffins breed in Labrador (and, recently, in northern Quebec and south-western Baffin Island where, hopefully, they will become established), in west Greenland, on Iceland, Jan Mayen, Bear Island, Svalbard and Novaya Zemlya, and in northern Fennoscandia. In winter the birds move to a broad band of the North Atlantic from Newfoundland and southern Greenland to the British Isles, and also to the North Sea and the coasts of southern Iberia and North Africa.

Horned Puffin *Fratercula corniculata*
Breeding adults are very similar to Atlantic Puffins, differing in the bill and facial features. The bill is much larger. The distal third is red, the basal two-thirds yellow. The eye is set in a narrow bare red ring, from the top of which a black horn of skin projects upwards. There are smaller horns below the eye. In winter adults are as breeding, but the facial disc is grey, darker in front of the eye, and the horn is lost or vestigial. The basal upper mandible plate is lost so the base looks very constricted. The distal third is dull red, the basal section dull brown.

Horned Puffins breed on Wrangel Island, in eastern Chukotka, on the shores of the Sea of Okhotsk, on Sakhalin and the Commander and Kuril Islands, on the Aleutian Islands and the Alaska Peninsula, Bering Sea islands and the west coast of Alaska. In winter the birds are found in a broad band of the north Pacific from northern Japan to the central United States.

Tufted Puffin *Fratercula cirrhata*
Breeding adults are very dark brownish-black overall, apart from a white facial disk. From above and behind the eyes two long (up to 7cm) sulphur-yellow plumes curve back over the nape. The lower mandible of the large bill is red. The distal two-thirds of the upper mandible is also red, the basal third dull yellow. In winter adults are as breeding, but the underparts are paler and often have a few white spots. The facial disk is uniformly dark brown. The exotic plumes are lost, but the feathering behind the eye is yellow-brown. Loss of the basal plate means that the bill is constricted, but much less so than in the Horned Puffin. The distal two-thirds are dull orange-yellow, the basal third dark brown.

Tufted Puffins breed in eastern Chukotka and on Kamchatka, at limited sites in the Sea of Okhotsk and on Sakhalin, the Commander and Aleutian Islands, on Bering Sea islands, western Alaska, the Alaska Peninsula, and on south and south-east Alaskan coasts. In winter the birds are seen in a broad band across the North Pacific from northern Japan to the western United States.

a Atlantic Puffin with prey.
b Atlantic Puffin.
c Horned Puffin.
d Horned Puffin pair with nesting material.
e Tufted Puffin pair.
f Tufted Puffin.

b

d

f

Owls

One of the most recognisable of bird families though, paradoxically, many people have never seen one because of their nocturnal habits. There are diurnal owls, but most are night hunters.

Owls are squat birds with wide heads and huge eyes. Head size and shape is an adaptation to help improve the acuity of both sight and hearing. In cross section, the owl's eyes are cylindrical (as opposed to spherical as in mammals such as humans). This allows a larger pupil and lens, and so improves light capture. The eyes are also mounted frontally to improve binocular vision. The disadvantage of these adaptations is that they reduce the field of view (to about 110° compared to 180° in humans). To compensate, owls have highly flexible necks that allow the head to be turned almost 270°. Although owl vision is only 2–3 times better than that of humans, this improvement is highly significant at night.

The wide head aids hearing. A notable feature of owls is the facial disc, a dish of stiff feathers (much more developed in nocturnal species than in diurnal owls). The feathers channel sound to the ears, much as parabolic dishes aid sound reception for human observers. The ears are far apart, allowing the bird to detect small differences in the arrival time of a sound across the head. This gives information on the location of the noise in the horizontal plane. Some owls also have asymmetric ears (usually the right ear set high, the left set lower) providing information in the vertical plane. Those owls that do not have such asymmetric ears can move ear-flaps that alter the size and shape of the ear to obtain the same information. So good are owls at locating sound that in experiments they have been able to catch prey in complete darkness, hunting by sound alone. Indeed, Snowy Owls (and others, such as the Great Grey Owl) can hunt rodents through several centimetres of snow with no visual clues. One further adaptation is that the ears are particularly sensitive to high frequency sounds, such as the rustling of a rodent making its way through dead leaves.→

Snowy Owl *Bubo scandiacus*
Status: CITES: Appendix II.
Magnificent birds that provide one of the great sights of the Arctic. Breeding males are white, occasionally with a few brown spots on the crown and upperparts (Photo 183a, standing bird). Breeding females have a white head, but the rest of the bird is white with, usually heavy, dark brown barring (Photos 183b and f). The eyes are golden-yellow, set in an incomplete facial disc. Breeding in continuous daytime, Snowies would appear to require a much less pronounced disc and it is also the case that in winter many Snowies move south and then hunt diurnally. But some overwinter in the Arctic and must hunt in complete darkness. And they survive, so their hearing is clearly adequate for detect non-hibernating rodents going about their daily lives beneath a blanket of snow.

While most owl species lay a reasonably well-defined number of eggs, Snowies are renowned for the variability of their clutch. In 'lemming years', when the rodent population explodes, the female owl may lay up to 14 eggs, the number determined less by prey availability but the time the male, who does all the hunting to feed the hatchlings, can spend flying, catching and returning with prey. In such lemming years, the owl population can itself explode, but many owlets will not survive the following season when rodent numbers crash. In the latter years the female Snowy may lay only two eggs with the male augmenting rodents with birds such as grouse. One interesting aspect of owl nesting is that other species sometimes breed close to the owls, trading the possibility of their chicks being predated against the reduction in losses to other predators which the Snowies see off (though they have their own problems occasionally – Photo 183d and e, a Long-tailed Skua chasing off an owl flying too close to its nest. Most famously Red-breasted Geese breed in close association, but as Photo 183a (Wrangel Island) shows, Common Eider will also take advantage of the owls 'protection'.

Breeds in north Greenland, northern Fennoscandia, northern Russia (including Novaya Zemlya, Severnaya Zemlya and Wrangel Island), west and north Alaska and northern Canada including the southern Arctic islands, but the distribution is dependent on prey numbers: Snowy Owls have bred on Iceland and Jan Mayen, probably on Svalbard and the other Russian Arctic islands, and on Canada's Arctic islands north to Ellesmere. Essentially resident, moving south if prey numbers decline.

Short-eared Owl *Asio flammeus*
Status: CITES: Appendix II.

Breeding males have buff and white facial discs, with a black bill and yellow eyes. The underparts are pale buff, paler still on the belly, the whole streaked with dark brown, the streaking heavier on the breast and throat. The upperparts are tawny-buff, heavily streaked dark brown. Breeding females are similar, but deeper buff and, in general, are more heavily marked. Photo 185c, in which the owl looks very similar to a Russian doll, illustrates the ease with which the owl can turn its head through 180° to look directly over its back.

In studies of the owl's diet mammals constituted around 95% of the biomass, with rodents and shrews providing the bulk of that percentage, though the owls also take stoats and weasels. Birds, amphibians, reptiles and insects provide the remaining 5%. Arctic breeding Short-eareds are invariably ground nesters (Photo 185ed Siberian shrub-tundra), Short-eareds being one of few owls that make a 'proper' nest. In it the female lays 2-13 eggs, the species reacting in a similar fashion to Snowy Owls in lemming years.

Short-eared Owl breeds on Iceland, in northern Fennoscandia and across Russia to Kamchatka, though rarely to the north coast. Breeds in Alaska, the Yukon and North-West Territories, around Hudson Bay, in Labrador and on southern Baffin Island. At all times the actual distribution of the owls is dependent on prey density, but northern owls do move south to north-west Europe, central Asia and continental United States.

→Owls are densely feathered, the feathers having soft fringes so their flight is quiet, the normal hunting technique being a pounce onto prey from a hovering flight. Tundra species, which hunt by day, have long wings that allow an efficient quartering flight as they visually search for prey. The hooked bill is small and often almost completely hidden in the feathers. The talons are sharp and the outer toe can be reversed, improving both capture area and grip. Owls usually carry their prey in their bills, though occasionally larger prey may be carried in the feet (in the manner of raptors). The prey is ingested whole, with the indigestible portions – fur, bones etc. – being periodically disgorged. The examination of pellets allows a study of the owl's diet. In times of abundant prey owls may cache food.

Despite the name of some owls (such as the Short-eared Owl), the prominent 'ear' tufts of these species have nothing to do with hearing or, indeed, ears. The tufts are for communication between conspecifics.

Not surprisingly for essentially nocturnal creatures, plumage colour is generally not so important for owls, with many forest-dwelling species having a cryptic plumage that allows the bird to rest undisturbed during the day. Some species that inhabit a latitudinal range encompassing both deciduous and coniferous forests actually change their plumage base colour from brown in the south to grey in the north. In general plumages are similar for the sexes and for both summer and winter.

Because sight is less important for territorial and mating purposes, sound has replaced it as the primary means of communication in owls. Pair formation is therefore based on the male hooting his territorial claim. For the two true Arctic breeders, visual displays are more appropriate, male Snowy Owls having a flight with the wings raised in a 'V', and a wing-raising ground display. Male Short-eared Owls clap their wings. In general pairs are monogamous and seasonal, though males may be polygamous in good prey years. Owls breed early in the season so that young rodents are plentiful when their chicks are learning to hunt, and when moulting adults are least efficient hunters. The boreal species nest in tree holes, but the two true Arctic owls nest on the ground. The number of eggs is highly dependent on prey numbers and can be as low as 1 or 2 or as high as 10–13. The female incubates while the male feeds her. The owlets are nidicolous and altricial, and are dependent on their parents for a relatively long period. The summer and winter plumages of the owls described here are similar.

c

e

(Northern) Hawk Owl *Surnia ulula*

Status: CITES: Appendix II.

Adults have white facial discs ringed black. The eyes are yellow. The underparts are white, finely barred brown (very similar to the Goshawk). The upperparts are grey-brown, mottled white. The long tail is dark grey-brown with delicate white barring.

A boreal species, but seen to the treeline, Hawk Owls breed in Fennoscandia and across Russia to Kamchatka. Breeds in central Alaska (but is uncommon in west and south) and across Canada to the Atlantic coast. Resident and irruptive, as for other owl species.

Tengmalm's Owl (Boreal Owl) *Aegolius funereus*

Status: CITES: Appendix II.

Adults have white facial discs ringed black, similar to Hawk Owl, but usually exhibiting a more 'startled' look. The underparts are white, heavily smudged dark brown. The upperparts are dark brown spotted white, the spotting heavier on the head. The tail is dark brown, finely barred lighter brown. As with other owls, Tengmalm's will cache prey if it has had a successful day. This is a particularly good strategy in winter as the owl is less able to hunt through the snow than other owls. Tengmalm's caches are usually in tree holes. In winter the prey freezes, useful as a storage method, keeping the food from decomposing, but difficult for the owl when it comes in search of a meal. To overcome the problem the owl will 'brood' the prey item, just as it would its eggs, thawing it to the point where it can be ingested.

A boreal species, but seen to the treeline, Tengmalm's Owls breed in Fennoscandia and across Russia to Kamchatka. Breeds in central Alaska (but is rare in the west and uncommon in the south) and across Canada to the Atlantic coast. In general seen further south than the Hawk Owl. Resident and irruptive, as for other owl species.

Great Horned Owl *Bubo virgnianus*

Status: CITES: Appendix II.

Adults are variable across the range. In central and eastern Canada the birds have rufous-brown facial discs, ringed black. The underparts are buff, heavily barred dark brown or dark grey-brown. The upperparts are grey-brown, mottled dark brown and white. In western Canada the owls are paler, the facial disc pale grey, the underparts pale grey barred darker, the upperparts mottled grey and white. In Alaska the birds are much darker, with brown facial discs, the underparts being dark buff, sometimes even pale chestnut, with dark brown barring, and the upperparts mottled dark brown, brown and white. The owl has prominent ear tufts and large yellow eyes.

A boreal species, but seen to the treeline, Great Horned Owls breed in central and southern Alaska (but are uncommon in the west) and across Canada. Resident and irruptive, as for other owl species.

Great Grey Owl (Great Gray Owl) *Strix nebulosa*

Status: CITES: Appendix II.

Adult nominate Eurasian owls have highly distinctive facial discs, with pale grey 'commas' on the inner side of the yellow eyes, the lower sections of which emphasise the yellow bill. The feathering around the bill is black. The remaining facial disc comprises concentric three-quarter circles alternately pale and dark brown. The underparts are cream/buff streaked dark brown, the upperparts having the same base colour, but the streaking being much heavier. The tail is pale grey, browner towards the tip and with very indistinct bands. North American owls, *S. n. lapponica*, are darker than their Eurasian cousins.

A boreal species seen in coniferous and mixed woodland, and often preferring mountain areas. Breeds in northern Fennoscandia and across Russia, but only to western Chukotka, and absent from Kamchatka. Breeds from central Alaska across Canada to the southern shores of Hudson Bay and southern Quebec. Resident and irruptive, as for other owl species.

a, b Northern Hawk Owls.
c Tengmalm's Owl.
d Great Horned Owl.
e, f Great Grey Owl.

b

d

f

Kingfishers and Woodpeckers

Most kingfishers live in the tropics, with only one bird breeding in the Arctic as defined in this book, and then only at the fringe. As with its more southerly cousins, the Belted Kingfisher of the Nearctic is superbly adapted for its feeding strategy of long periods of patience interrupted by short bursts of activity. The Belted Kingfisher (*Megaceryle alcyon*) breeds in central Alaska, central Yukon and North-West Territories, and on the southern edge of Hudson Bay. In winter the birds move to south-eastern Alaska and southern USA. Photo 189a is of a breeding male. Breeding females have a second, narrower chestnut breast-band and chestnut striping along the upper flanks. Summer and winter plumages are similar.

Woodpeckers are boreal species and so technically non-Arctic. Several species may occasionally be seen at the treeline, but only three are regularly seen at the northern extremity of the forest. The Three-toed Woodpecker (*Picoides tridactylus* – Photo 189c) breeds to the treeline across both the Palearctic and Nearctic. In the western Palearctic and over much of the Nearctic the birds are resident, but in eastern Russia the birds move south to avoid the harsh Siberian winter. It is likely that some Nearctic birds also move south in winter. The odd name reflects the fact that unlike most woodpeckers, which have four toes arranged as two pairs, one pair pointing forward, the other backward to allow efficient tree climbing and provides a solid anchor for pecking, the Three-toed has only three toes. The Northern Flicker (*Colaptes auratus*) breeds across North America. There are two major forms (and many subspecies of them) distinguished by the shaft colour of the remiges, which give the yellow- and red-shafted forms differing colours in the underwing. The northerly birds are *C. a. aurates*, yellow-shafted birds (Photo 189b). These northerly birds head south for the winter. The Lesser Spotted Woodpecker (*Dendrocopos minor*) breeds across the Palearctic to the west coast of the Sea of Okhotsk and on Kamchatka. Photo 189d is of a female.

Larks

Larks are ground–dwelling passerines that walk (rarely running and almost never hopping) in search of seeds and insects. Because of their habitat and lifestyle they are, in general, cryptic brown, and have long legs, and a long hind claw to aid standing. Only one species breeds in the Arctic. The Shore (or Horned) Lark is a typical lark: flocks form for migration and wintering. Males hold territories and entice females by calling. The pair bond is monogamous and seasonal. The nest holds 3–4 eggs. Incubation is by the female, but both parents feed chicks.

Shore Lark (Horned Lark) *Eremophilia alpestris*
Breeding males are very distinctive. The head is pale yellow, with a black horseshoe through the eyes and bill, a black necklace and a brown crown and nape. At the front of the crown there is a black bar that ends in black tufts (the 'horns' of the North American name). These are raised as a threat to males intruding on a occupier's territory. The underparts are white, the flanks smudged pinkish-brown. The upperparts are mottled pink, brown, dark brown and grey. Breeding females do not have horns. The plumage is as the male, but is duller overall and lacks the black horseshoe and necklace, these being pale grey. Female summer and winter plumages are similar. Males lose their horns in winter, and are then similar to the female. Shore Larks have a complex taxonomy, with more than 40 subspecies having been described. In the Palearctic the bird is a northern dweller only, with southern habitats being occupied by the Eurasian Skylark (*Alauda arvensis*). Shore (Horned) Larks are the only Nearctic larks, the species breeding in both northern and southern habitats.

Shore Larks breed in Fennoscandia, on Novaya Zemlya's southern island and north to the coast across Russia to the Kolyma delta. Breeds throughout Alaska and northern Canada, including Arctic islands to Devon Island. In winter Palearctic birds are seen in south-eastern Europe and southern Russia, while Nearctic birds fly south to join resident southern birds.

e Female Shore Lark of the nominate race photographed in northern Manitoba.
f Male Shore Lark of the race E. a. hoyti *photographed in Nunavut.*
g Incubating female Shore Lark of the northern Eurasian race E. a. flava.

Swallows and Martins

Among the most popular of birds as their arrival in northern Europe heralds the arrival of summer, the birds would be even more popular with the Arctic traveller if only they occurred in greater numbers, as they take insects on the wing and so reduce – but sadly only by a minimal amount – the number of mosquitoes.

The birds have long, narrow, pointed wings and deeply forked tails: the high aspect-ratio wings make the birds fast, but limit manoeuvrability, the forked tail restoring that so the birds are quick and agile. In general martins have shorter tails than swallows. The specialised diet of winged insects means that northern species move south in winter to where insects are found.

The birds have limited displays, male Sand Martins holding a burrow that a female might favour. The pair-bond is monogamous and seasonal, but males will attempt to mate other females. Swallow nests are often made of mud pellets, but they may be more conventional if a suitable crevice is found. 4–5 eggs are incubated by the female only, and she also takes responsibility for brooding the chicks, though both parents feed them. The summer and winter plumages of the species described here are similar.

Tree Swallow *Tachycineta bicolor*
Breeding males have a shiny blue crown to eye level. The throat and underparts are white. The upperparts are blue, with a sheen that can sometimes appear greenish (Photo 191a). Adult females are as males, but the crown and upperparts are a dingy grey-brown. Tree Swallows differ from most other members of the family by feeding on vegetation and berries in addition to insects. They take these at all times, but particularly in bad weather.

Breeds throughout Alaska (but rare in the north), central Yukon and NWT, around the southern shores of Hudson Bay, in southern Quebec and in eastern Labrador. In winter the birds fly to southern USA, Mexico, and the east coast of Central America.

Cliff Swallow *Hirundo pyrrhonota*
Breeding adults have blue crowns, white foreheads and red cheeks, a black throat patch and a white nape linking to a white collar on the upper breast. The rest of the underparts are white or pale grey, with pink smudging on the flanks. The upperwing is blue-grey, the underwing pale grey (Photo 191b). The blue-grey tail is notched rather than forked. Cliff Swallows nest on the ferry that makes a triangular crossing of the Mackenzie and Arctic Red rivers to take the Dempster Highway north towards Inuvik in NWT (Photo 191d), parents birds needing to follow the ferry in order to feed their chicks.

An Arctic-fringe species, breeding in central Alaska, but uncommon elsewhere in the state. Also breeds in northern Yukon and western NWT (though not to the north coast), then southerly across Canada. Winters in southern South America.

Violet-Green Swallow *Tachycineta thalassina*
Breeding males are extremely attractive, with emerald-green crowns and backs, white cheeks, throat and underparts and dark grey-brown, purple-sheened upperwings. (Photo 191c). Females are patterned as males, but much duller.

An Arctic-fringe species, breeding in central and southern Alaska and in Yukon (but rarely to the north coast), and then more southerly in Canada and the USA. Winters in the southern USA and Mexico.

Sand Martin (Bank Swallow) *Riparia riparia*
Breeding adults have mid-brown upperparts, the forehead slightly paler. The brown of the crown extends across the cheeks (usually darker around the eye) and as a broad band around the upper breast. The rest of the underparts are white apart from the brown undertail. The tail is forked, but not deeply. (Photos 191e and f).

Sand Martin breeds in northern Fennoscandia, around the White Sea and then more southerly across Russia to Kamchatka. Breeds in central and western Alaska (though not common in the west), near the Mackenzie delta, then more southerly to southern shores of Hudson Bay and in southern Quebec and Labrador.

Pipits and Wagtails

The members of the Motacillidae are seed- and insect-eating passerines, which can be usefully divided into two groups, the wagtails, which are generally boldly coloured with greys, black, white and yellows, and the more cryptic pipits. The wagtails have long tails, the name deriving from the endearing habit of tail-wagging, which the birds do almost continuously while on the ground. The pipits are renowned for male display songs, the birds flying high (to 100m: out of sight to human observers) and delivering territorial claims while apparently hanging motionless (by flying into the wind). The flights can also be long-lasting, occasionally taking several hours, though they are often much shorter, and end with the bird 'parachuting' back to earth, this involving fluttering wings and a near-vertical descent.

A male pipit's song defines its territory and encourages females. Wagtails also have a territorial song, this usually starting from a perch. Male wagtails point their bills upwards to expose the breast patterns, and may also fan their tails. The pair-bond is monogamous and seasonal. The nest is a cup of vegetation well-hidden on the ground. 2–5 eggs are laid. These are generally incubated by both birds in the wagtails, the female only in the pipits, but there are exceptions. The chicks are nidicolous and altricial and are cared for and fed by both parents.

Red-throated Pipit *Anthus cervinus*
Breeding males have pale orange-buff heads and breasts, the crowns streaked brown. The rest of the underparts are white with black streaking. The upperparts are mottled tawny brown, black and white – the typical pipit cryptic plumage (Photo 193a). Females are as males, but with paler heads and breasts. In winter both sexes have paler heads and breasts.

Red-throated Pipits breed in northern Fennoscandia and across Russia to Chukotka and Kamchatka, to the north coast except on the Taimyr Peninsula. Also breeds in small numbers in western Alaska. In winter the birds are seen in central Africa and south-east Asia.

Pechora Pipit *Anthus gustavi*
Breeding adults have black and buff streaked crown and upper head, the lower head being pale buff or white. Breast and flanks are buff/yellow-buff, becoming paler on remaining underparts. Breast and flanks heavily streaked dark brown. The upperparts are typically 'pipit', mottled black, brown and buff (Photo 193b). In winter the birds are duller overall.

Pechora Pipits breed from the Pechora River eastwards to Chukotka, but to the north coast only in Chukotka. Also breeds throughout Kamchatka and on the Commander Islands. In winter the birds fly to Indonesia and the Philippines.

Meadow Pipit *Anthus pratensis*
Status: IUCN: Near threatened (habitat loss due to intensive agriculture).
The archetypal pipit, adults having crown, upper head and upperparts which are mid-brown, sometimes olive-brown, heavily streaked with dark brown or black. The throat and underparts are cream or buff, often orange-buff, on the flanks, with dark brown streaking on the breast and flanks (Photos 193c and d). Summer and winter plumages are similar.

Meadow Pipits breed in south-east Greenland, on Iceland and Jan Mayen, in northern Scandinavia and Russia east to the shores of the White Sea. In winter the birds fly to southern Europe, north Africa, the Middle East and central Asia.

Buff-breasted Pipit (American Pipit) *Anthus rubescens*
Breeding adults have grey-brown heads with a distinct buff supercilium. The throat and underparts are white, buff or yellowish-buff on the flanks. The breast and flanks are heavily streaked dark brownish black. The upperparts are less heavily streaked than other pipits, the streaking very fine. Several subspecies – Photos 193e (*A. r. pacificus* – northern North America) and 193f (*A. r. japonicus* – east Siberia). In winter the birds are duller overall.

Buff-breasted Pipits breed in Alaska, across mainland Canada to the north coast and on southern Arctic islands from Banks to Baffin. Also breeds in north-west and west central Greenland and in Chukotka and Kamchatka. In winter Nearctic and Greenlandic birds fly to the southern United States and Central America, while Asian birds head for southern Asia.

White Wagtail *Motacilla alba*
Adult males have white heads with a black crown. The throat, lower cheek and upper breast are black. The rest of the underparts are white, smudged dark grey on the lower flanks. The upper body is grey, the tail having a black centre and white outer feathers (Photo 195a). Adult females are as males, but the pattern is less striking and smudged, the upperparts darker grey. In winter plumage males and females are similar, and as the summer female.

Breeds in south-east Greenland, Iceland, Jan Mayen, northern Fennoscandia, and across Russia to Kamchatka, north to the coast except on Taimyr, and on Novaya Zemlya's southern island and Wrangel. Also breeds irregularly in north-west Alaska. In winter the birds are seen in north and central Africa, the Middle East, and south and south-east Asia.

Yellow Wagtail *Motacilla flava*
The subspecies taxonomy of Yellow Wagtails is complex and confusing, and it remains to be seen whether the decision of the AOU in early 2016 to give full species to the American Yellow Wagtail (*M. tschutschensis*) on the basis of genetic studies helps. Essentially all birds have pale yellow throat, the rest of the underparts yellow with olive smudging on the flanks. The upperbody is olive green, the central tail dark grey, the outer feathers white. Head colour and pattern varies with subspecies. In northern Scandinavia and across much of Russia *M. f. thunbergi* has a slate grey crown but otherwise black head, but white throat (Photo 195c). Photo 195b is of a Kamchatka wagtail. Formerly a subspecies of flava, *M. f. simillima*, it is likely now to be considered a subspecies of the Alaskan species which it closely resembles. The obvious difference from *thunbergi*, the distinct supercilium, is shared by many *flava* subspecies including those of northern Siberia.

Yellow Wagtail breeds in northern Scandinavia and across Russia to Kamchatka, north to the coast except on Taimyr. The American species breeds on the west and north coasts of Alaska. In winter the birds head for central and southern Africa, India and south-east Asia.

Waxwings

Waxwings are named after the wax-like red 'droplets' found at the tips of the secondary feathers (and, occasionally, the tertials and tail feathers) of two of the three species in the family. The 'droplets' (which are, of course, part of the feather structure) derive their colour from carotenoids in the birds' principle food, fruit. The size and number of the 'droplets' increases with age, and are an indication of mating fitness. The yellow wing feather-tips and tail-band also result from carotenoids and are further fitness indicators.

The birds are important seed-dispersers as they are entirely frugivorous, consuming huge quantities daily, up to twice their body weight. The birds are non-territorial even in the breeding season (a consequence of the transitory nature of fruiting). The birds are much more gregarious in winter, a rarity among non-marine species.

(Bohemian) Waxwing *Bombycilla garrulus*
Handsome birds. Breeding adults are pinkish buff overall, paler below than above, with crest at the back of the crown, black mask and throat. Plumages of sexes similar, with summer and winter plumages also similar (Photo 195d). Waxwings breed late, a response to the seasonal nature of fruit production. As autumn approaches, if stocks are poor the birds may be forced to seek alternative foods and this can result in irruptions, large flocks of waxwings being seen well south of their usual range and in areas from which they are usually absent. Being frugivores (though the birds do take insects), waxwings can find themselves feeding on fermenting fruit as autumn approaches. Studies indicate that the birds have evolved a relatively high capacity to metabolise ethanol, which makes them more alcohol-tolerant than other species, but despite being able to 'hold their drink' waxwings can become a little drunk and disorderly if fermented berries form a high proportion of their diet.

An Arctic-fringe species. Breeds in southern Fennoscandia, around Russia's White Sea coast, but then more southerly. Breeds in central Alaska, but rare elsewhere in the State, in northern Yukon and north-western North West Territories, but then more southerly to the south-west shores of Hudson Bay. In winter the birds are seen in southern Scandinavia and eastern Europe, southern Canada and northern continental United States.

Thrushes and Chats

The Turdidae are a large and widespread family of birds, with representatives on every continent apart from Antarctica, and which includes some of the best-known species to northern peoples – the Eurasian Robin (*Erithacus rubecula*), American Robin, Eurasian Blackbird (*Turdus merula*) and the bluebirds. It is thought that familiarity with remembered species led early European settlers in America to name the American Robin for its European cousin: the two species have rather more than their red breasts in common, as the widespread American species was also a harbinger of spring for northern settlers in the same way as the territorial call of the European species heralds spring for many. It is also likely that the numerous American blackbirds received their names because of the superficial resemblance to their European cousins. The family is, however, diverse, the many species having few characteristics in common. The birds are principally invertebrate feeders, supplementing this with fruit. Many species are excellent singers (the Nightingale *Luscinia megarhynchos* being famous for its voice), males calling to establish territories and for pair formation. Pair-bonds are usually monogamous and seasonal. Usually 4–6 eggs are laid in a well-built nest and normally incubated by the female only, though both birds usually contribute to the care of the chicks. The summer and winter plumages of the species below are similar.

Redwing *Turdus iliacus*
Status: IUCN: Near threatened (due to climate change and illegal trapping on migration). Breeding adults have a dark brown or grey-brown crown, a white supercilium, dark brown lores and cheeks and a white throat finely streaked brown. The upperparts are grey-brown, the flight feathers darker. The underparts are white or pale cream, heavily streaked dark brown on the breast and flanks. The undertail is grey-brown. The underwing has rusty coverts, merging with rusty flank patches, these giving the bird its name.

Redwings breed on Iceland, in northern Fennoscandia and across Russia (though not in Chukotka or Kamchatka). Some Icelandic birds are resident, but in general the birds move to northern Europe for the winter.

Fieldfare *Turdus pilaris*
Breeding males are highly distinctive. The head is blue-grey apart from a white supercilium and throat, the latter heavily streaked dark brown or black and tinged orange. The breast is orange-buff, the rest of the underparts white, though the flanks are tinged orange-buff. The breast and flanks are heavily streaked dark brown, the markings on the flanks being arrowheads. The belly and vent are unmarked. The mantle is brown, the back and uppertail pale grey. The long tail is dark grey. The upperwings are grey and grey-brown, the underwing pale grey and white. Breeding females are as males, but duller overall.

Fieldfares nest colonially, a very unusual habit among thrushes. The reason is a highly coordinated defensive system against predators. When a potential predator is spotted the birds fly towards it. Each in turn then dives at the intruder, letting loose a bomb of excrement. The bird's aim is usually good, with the intruder rapidly becoming peppered with spots of excrement. On a predatory bird such spotting can be dangerous as it affects the aerodynamics of the feathers, and on terrestrial predators it can require a lot of grooming to remove. Not surprisingly the strategy is extremely effective at persuading predators to retreat. It is so effective that other birds choose to nest close to Fieldfares and so benefit from the defence it brings. Most interestingly, a study in Scandinavia showed that Fieldfares and Merlins nested close, the falcons chasing corvids away and so reinforcing the Fieldfares anti-predator strategy, and that both Bramblings and Chaffinches (*Fringilla coelebs*) have higher chick survival rates if they nest close to the Fieldfares/Merlins.

Fieldfare breeds in northern Fennoscandia and in western and central Russia, though not to the north coast. Has bred in Iceland, but not established. In winter some Scandinavian birds fly to Iceland (where some may remain to breed), others, and birds breeding in Russia, fly to southern Europe, southern Russia and central Asia.

a, b Redwings. The male in 'a' is singing his territorial claim.
c, d Fieldfares.

American Robin *Turdus migratorius*

American Robins, the USA's 'favourite songbird' is the largest Nearctic thrush. This enthusiasm is doubtless aided by the fact that the birds are often seen in suburban areas, being just as happy there as in rural landscapes. They are also very widespread, northern species being true Arctic dwellers, while other individuals do equally well in the far hotter climates of the southern states. Breeding adults are striking birds, with dark grey heads and white-streaked throats. The breast, belly and flanks are red, the vent and undertail white. The upperparts are grey, the tail dark grey with white tips to the outer feathers. There are seven subspecies across the continent, the differences between them being subtle coloration changes. The two northern subspecies, one occupying Alaska and western Canada, the other eastern Canada, are the darkest robins.

American Robin breeds throughout North America, including Alaska (though uncommon in the north), the Yukon and much of North-West Territories. The range is more southerly in eastern Canada, but includes the southern shore of Hudson Bay and southern Quebec and Labrador. In winter the birds are seen in southern United States and Central America.

Bluethroat *Luscinia svevica*

A most beautiful bird, but one that is often elusive. Breeding males have a grey-brown crown, forehead and nape, a white supercilium, chestnut upper cheeks and pale grey lower cheeks, the latter blending into the nape. The throat and upper breast are metallic blue, enclosing an orange-chestnut patch. The breast is bordered by bands of black and orange-chestnut. The rest of the underparts are pale buff or cream. The upperparts are grey-brown. Breeding females are as males, but the throat and breast are white with a chestnut breast band. Some females show two breast bands, one blue, the other chestnut. The described birds are as seen in Scandinavia. The taxonomy of the species is complex with a number of subspecies being described, these varying subtly in colour, the differences not helped by birds showing a marked degree of individual coloration within the subspecies.

Bluethroats breed in northern Fennoscandia and across Russia to Chukotka, but rarely to the north coast. Also breeds in small numbers in western Alaska (where the arrival of the birds brings excitement and crowds (a relative term, there are never crowds in the accepted sense, but numbers do increase) to Nome and the Seward Peninsula. In winter the birds move to the coasts of the Mediterranean and to central Africa.

Naumann's Thrush *Turdus naumanni*

Breeding males have a buff-speckled dark brown crown and white supercilium. The face is white, with brown cheek patches. There is a dark brown breast-band. The rest of the underparts are white, heavily marked with dark brown arrowheads. The upperparts are chestnut, mottled with black, the wings more evenly chestnut. Breeding females are as males, but duller overall. The bird described here is the subspecies *T. n. eunomus*, which breeds to the north of the nominate. *T. n. eunomus* is markedly different from the nominate and is often called the Dusky Thrush, particularly in North America where it is a casual visitor.

T. n. eunomus breeds in Russia from the Yenisey to central Chukotka and throughout Kamchatka. In winter the birds fly to south-east Asia.

a American Robin, Great Slave Lake, NWT, Canada.
b American Robin, Seward Peninsula, west Alaska.
c, d Male Bluethroats, illustrating the colour difference which can be seen in individuals within areas. Both photographs were taken in northern Norway.
e, f Naumann's Thrush T. n. eunomus.

Grey-cheeked Thrush (Gray-cheeked Thrush) *Catharus minimus*

Breeding adults have olive-brown or brown-grey heads with grey-buff cheeks, these occasionally finely streaked paler. The throat is white, separated from the cheeks by a dark stripe. The breast is buff, heavily spotted dark brown. The rest of the underparts are white, smudged brownish-grey on the flanks and belly. The upperparts and tail are uniformly olive-brown, the flight feathers darker.

Grey-cheeked Thrush breeds throughout northern Alaska and the Yukon, then more southerly to the southern shores of Hudson Bay and in most of Quebec and Labrador. Also breeds in small numbers in Chukotka. In winter the birds are seen in northern South America.

Varied Thrush *Ixoreus naevius*

Extremely attractive birds. Breeding males have a lilac crown and nape, an orange-brown supercilium, a black forehead and a broad black eye-stripe. The throat and upper breast are orange-brown, becoming paler towards the vent. The flanks are smudged lilac. The upperparts are lilac, the tail dark lilac-black. Breeding females are as male, but duller.

Varied Thrush breeds throughout Alaska (apart from the extreme north) and in the Yukon and western North-West Territories. In winter the birds move to the west coast of the United States.

Northern Wheatear *Oenanthe oenanthe*

Breeding males have a pale blue-grey crown and nape, a white forehead and a distinct white supercilium. The throat is pale orange-buff, the breast paler, the rest of the underparts white. The upperparts are pale blue-grey, the central tail black, the outer feathers white. Adult females have a grey-brown crown, nape and upperparts. The underparts are usually a richer buff but they may be paler than the male. The upperwing is dark brown, the underwing as the male, but washed buff.

Northern Wheatears breed on both coasts of Greenland, but not to the extreme north on either, on Iceland, irregularly on Svalbard, and from northern Scandinavia to Chukotka, north to the coast in most areas, and on the southern island of Novaya Zemlya. Breeds in western Alaska, rarely in the northern State, in the Yukon. In the eastern Canada Wheatears are birds of the High Arctic, breeding on Baffin and Bylot Islands, Devon and Cornwallis islands and southern Ellesmere Island. They also breed on the mainland, in northern Quebec and Labrador. The nominate Wheatear, as described, breed across Eurasia. The birds of North America, Greenland and Iceland are a subspecies, *O. o. leucorhoa*, which tend to be larger and have underparts which are a richer buff. However, richer coloured specimens of the nominate are known, and as the illustrations on p185 indicate there is variation between individuals. There are also several extralimital subspecies.

Given the essentially circumpolar distribution of the Northern Wheatear it would be thought that in winter the birds would be found equally well-spread around the globe. But this is not the case, the entire population flying to central Africa. This curiously asymmetric winter distribution indicates the original breeding range of the bird – it was a northern European species that has spread east and west. This accounts for the Nearctic distribution, where the bird is absent from a huge central section of the continent.

a Varied Thrush.
b Grey-cheeked Thrush.
c, d Male and female nominate Northern Wheatears, photographed in northern Norway.
e Male O. o. leucorhoa, west Greenland. The bird has distinctly richer coloration, not only on the underparts, but on the upperparts.
f Female O. o. leucorhoa, west Greenland. This bird was the mate of 'e' and not only does it appear no richer in coloration than the female of 'd' it is actually less colourful.
g Male O. o. leucorhoa, Iceland. This bird is certainly no richer in coloration than 'c'.

b

d

g

Warblers

Only two species of warblers are northern breeders in the Palearctic, and only one of these is a true Arctic dweller. In each case both male and female, and summer and winter plumages are similar. Two further species, Middendorff's Grasshopper Warbler (*Locustella ochotensis* – Photo 203b) and the Lanceolated Warbler (*L. lanceolata*) may be seen in Kamchatka.

Arctic Warbler *Phylloscopus borealis*
Adults have dark olive crown, broad pale greenish-yellow supercilium and a dark olive-brown eye-stripe. Cheeks are olive-green, throat is pale olive, the rest of the underparts cream, the breast streaked olive (Photo 203a). Essentially boreal, but seen in forest-tundra. Breeds in northern Scandinavia and east to Chukotka, though infrequently to the north coast. Breeds on Kamchatka and the Commander islands. Has bred on Wrangel Island. Also breeds in western Alaska. In winter the birds are seen in the forests of south-east Asia.

Willow Warbler *Phylloscopus trochilus*
Patterned as Arctic Warbler but with grey-brown crown and upperparts, cream or pale yellow-cream supercilium, and grey-brown eye stripe. Most birds have pale underparts (Photo 203c), but some can be pale yellow (Photo 203d). Breeds in northern Scandinavia and east to Chukotka.Winters in sub-Saharan Africa, a very long flight for east Russian birds.

Tits and Chickadees

Tits are small passerines with pale heads beneath dark or coloured crowns. The bill is short and pointed, ideal for catching the birds' main prey of insects. Tit chicks are fed exclusively on insects and insect larvae (requiring upwards of 10,000 individual items before fledging). The Nearctic chickadees look very similar and have a similar diet. All the birds described here are boreal, though found to the treeline and are, consequently, Arctic-fringe species.

Siberian Tit (Grey-headed Chickadee) *Poecile cinctus*
Adults have dark brown crowns and nape. Remainng head is white apart from a black bib. The underparts are white, with rusty-brown breast sides, flanks, vent sides and undertail coverts (Photo 203e). Breeds in northern Scandinavia and east to Chukotka, but not to the north coast, and in Kamchatka. Breeds in small numbers in western and central Alaska. Resident or nomadic. Resident birds survive winter temperatures as low as -60°C. At temperatures of about -40°C the birds take shelter in a tree hole, tuck in their heads and feet and fluff out their feathers. As the temperature falls the bird drops its body temperature by up to 10°C, entering a state of torpor. In order to maintain normal body temperature and to build the resources necessary to survive periods of torpor, the bird must consume around 55–65% of its body weight daily, a quantity that must be foraged in a day as short as 4 hours.

Black-capped Chickadee *Poecile atricapillus*
Breeding adults have black caps and a large black bib. The rest of the face and underparts are white, with rusty smudging to the flanks and vent (Photo 203f). Breeds in Alaska, but uncommon except in the central State, and more southerly across Canada. Resident or nomadic.

Boreal Chickadee *Poecile hudsonicus*
Similar to the Siberian Tit, but with a paler cap, larger bib and a grey nape. In general the underparts are overall more rusty (Photo 203g). Breeds in central Alaska, but rare in the west. Also in Yukon, then more southerly across Canada to the southern shores of Hudson Bay, and in southern Quebec and Labrador. Resident or nomadic.

Willow Tit *Parus montanus*
Adults have black crowns and bibs, the rest of the head white. The underparts are white, tinged buff on the flanks and towards the vent (Photo 203h). Breeds in northern Scandinavia and east to Kamchatka (birds there are a sub-species that are very pale apart from the crown and bib).

Corvids

The crows are among the most familiar of birds: large, noisy and opportunistic, they are well-known to people throughout the world. They are also renowned for their intelligence, solving puzzles and able to use simple tools. Experiments also indicate that crows have excellent memories, being able to recall the location of thousands of cached food items (and to know which sort of food is in which location), and to show evidence of forward planning based on past events, which implies self-awareness. In controlled experiments (and by the personal observations of birdwatchers everywhere), crows that watched others cache food and then retrieved it for himself (i.e. stole it) always ensured that no other bird was watching when he cached his own food, changing the location if he thought he had been overlooked. This implies the crow was capable of understanding the possible actions of others based on his own behaviour. This 'theory of mind' is something that human children do not develop until they are about three years old, and is linked to self-awareness.

Of the species listed below only the Raven is a truly Arctic species, the remainder occurring to the treeline, though some are occasionally seen on the tundra.→

(Northern) Raven *Corvus corax*
Adults are glossy black with a purple, purplish–blue, or even purplish–green sheen. The Raven's wedge-shaped tail is diagnostic. The Inuit claim that the reason the Raven is black dates from a time when all birds were white. One day a Raven and a Loon (diver) agreed to draw on each others feathers so they would be distinctive. The Raven went first, but his efforts so annoyed the Loon that it spat fish oil all over him and Ravens have been black since that day. In retaliation the Raven beat the Loon, so badly that it could barely walk, and loons were never able to walk properly again. For the Inuit the croaking call of the Raven is *kak*. Their story is that the first Raven was actually a man. Before a hunting trip with friends, the man, nervous that blankets (*kak* in Inuktituk) would be forgotten, repeatedly told his companions to look for them. So fed up did they become that they deliberately forgot the blankets and sent him back to fetch them. In his panic the man ran, then flew and finally became a Raven, still calling *kak*.

Ravens breed throughout Greenland, on Iceland, in northern Scandinavia and across Russia to Chukotka, north to the coast except on the Taimyr Peninsula. May also breed on Wrangel. Breeds throughout Alaska and across Canada, north to the coast and on the Arctic islands north to southern Ellesmere. Resident or dispersive.

Eurasian (Carrion) Crow *Corvus corone*
Although Carrion and Hooded crows are usually classified as subspecies, many authorities consider the two are separate species, and that classification is used here.

Carrion Crows are overall black with a purple, blue or green sheen. Although Carrion Crows may be seen in northern Europe, in general the farther north the traveller goes the more likely it is Hooded crows will be seen. However, in the Russia Arctic east of the Yenisey River and on Kamchatka it is Carrion Crow that will be seen. Resident or dispersive.

Hooded Crow *Corvus cornix*
Hooded Crows have black heads, neck, wings and tail. The rest of the plumage is grey apart from a black 'splash' on the breast.

Hooded Crows breed in Scandinavia and across Russia to the Yenisey; hybridisation occurs where the races meet. Resident or dispersive.

a A Raven playing. There seems no better explanation for the behaviour. The author once watched a Carrion Crow doing somersaults around a telephone wire attached to his house for no apparent reason other then it was fun.
b, c Ravens. The shots illustrate the tail shape and feather sheen of the birds.
d Carrion Crow chicks on an island off Kamchatka's east coast. The birds were being reared on a diet which included a great deal of Ancient Murrelet eggs.
e Hooded Crow.
f Hooded Crow taking advantage of low tide to fish in shallow water.

Grey Jay *Perisoreus canadensis*
Breeding adults have dark grey crowns and white or pale buff foreheads. The rest of the head is pale grey, as are the underparts (Photo 207a: Photo 207d shows a Jay sunbathing on a warm bed of the lichen *Cladonia rangiferina*). The tail feathers have white tips.

Breeds in central and western Alaska (though are uncommon in the latter), central Yukon and North-West Territories, then southerly to the southern shores of Hudson Bay, and in southern Quebec and Labrador. Resident or dispersive.

Siberian Jay *Perisoreus infaustus*
Breeding adults are attractive birds with a black crown, the rest of the head being grey. The breast is grey, merging to a pale orange-grey belly and vent. The mantle and back are grey, the rump rusty-brown. The tail is rusty-brown, apart from black central feathers. The wings are grey and rusty-brown (Photo 207b).

Breeds in northern Scandinavia and across Russia to the border of Chukotka. Resident or dispersive.

Spotted Nutcracker *Nucifraga caryocatactes*
Breeding adults are dark chocolate brown, the underparts and upperbody heavily spotted white. The vent is white and there is a white terminal band to the tail (Photo 207c).

A boreal species, found to the treeline. Breeds in southern Norway and Sweden, but more northerly across Russia to Kamchatka, though not to the north coast. Resident or dispersive, but also an irruptive migrant.

(Common) Magpie *Pica pica*
Breeding adults are black, with blue or green sheen, apart from a white belly and scapulars. The outer wing is black and white. The long tail has a green sheen. Photo 207e illustrates the opportunistic nature of corvids, the Magpie is searching for a meal in a discarded fast food wrapper in a town in Arctic Sweden.

Magpies breed in northern Scandinavia, then more southerly across Russia to the Sea of Okhotsk, and also in southern Chukotka and Kamchatka. Resident, with limited dispersal.

Black-billed Magpie *Pica hudsonia*
Breeding adults are as the Common Magpie (Photo 207f). Differences between the Nearctic and Palearctic birds are minimal and some authorities consider them conspecific.

Breeds in central and southern Alaska and in south-western Canada. Resident, with limited dispersal.

→Crows, particularly the Raven, have been incorporated into the myths of many peoples. They appear in the epic of Gilgamesh, the Bible, and the myths of the Romans, Greeks and Celts, and in Chinese and Japanese mythology: the Vikings associated the Raven with Odin. Ravens could be friendly, but were also associated with death, an association that probably arose from the birds feeding on battlefield dead, so they became birds of ill-omen. But for the native peoples of the Arctic they were usually seen in a positive light. They are at the heart of creation myths of Inuit, Chukchi and Koryak. Shamans often took Ravens as their familiars and the birds are seen on the totem poles of the native peoples of southern Alaska. The Inuit believe the Raven helps them with their hunt: if a raven flies overhead an Inuk will ask it if it has seen prey, and if it has the bird will dip its wing to point the direction.

The corvids include colonial nesters (e.g. the Rook *Corvus frugilegus*) and co-operative breeders, with last year's young helping to feed this year's brood (several Nearctic jays and crows), but the Arctic and Arctic-fringe breeders are solitary nesters. The pair-bond is monogamous (though male Common Magpies are promiscuous) and life-long or long-lived. Eggs are incubated by the female only (but both birds in the case of the Spotted Nutcracker), with the chicks cared for and fed by both parents. An old saying maintains that 'a crow lives three times longer than a man, and a raven lives three times longer than a crow', but in reality most crows do not make it past 10 years of age. Yet despite this, sexual maturity is not reached until the bird is 6 or 7 in some species, though an age of 2 or 3 is more usual. The plumages of male and female, and in summer and winter, are similar in the species described here.

b

d

f

New World (Wood-) Warblers

The wood-warblers of the Nearctic share features with Palearctic warblers, to which they are only distantly related, such as having straight, short, pointed bills ideal for taking insects (though most also take fruit in winter), but they are mostly flamboyant in plumage rather than drab. Indeed, they are the most colourful of the passerines of the Arctic fringe and Low Arctic, and a true delight. Male wood-warblers sing to establish a territory and attract a mate, although there are usually some visual displays as well. The pair-bond is monogamous and seasonal, but some males with exceptional territories may be polygamous. The 3–6 eggs are incubated by the female. Chicks are brooded by the female, but fed by both parents. Summer and winter plumages are similar for the species described here.

Yellow Warbler *Dendroica petechia*
Breeding males have yellow heads with an olive-yellow cap. The upperparts are olive-yellow, the wings darker. The underparts are yellow with red streaks (Photo 209a). Breeding females are as males, but lack the red streaking of the underparts (Photo 209b).

Breeds throughout Alaska (but rare in the north), Yukon and NWT, but rare in Nunavut, on the southern shores of Hudson Bay and in southern Quebec and Labrador. In winter northern birds join resident birds on Caribbean islands and in northern South America.

Yellow-rumped Warbler *Dendroica coronata*
Breeding males have a yellow skull-cap, the rest of the head black and grey, but with a white throat. The remaining plumage is a mix of grey, white and yellow (Photo 209c). Breeding females are a little duller overall and lack the yellow skull-cap. Northern (nominate) birds (as described and illustrated) are known as 'myrtle warblers' because of their exclusive winter diet of Myrica sp. fruits.

Breeding range is as Yellow Warbler, but less far north and west in Alaska. In winter the birds are seen in the southern United States, in Central America and on Caribbean islands.

Northern Waterthrush *Seiurus noveboracensis*
Breeding adults have dark olive-brown crowns and napes, a buff supercilium, dark olive-brown eye-stripe and paler cheek patch. The underparts are pale yellow with dark streaking to the breast and belly. The upperparts are uniformly dark olive-brown (Photo 209d).
Breeding range as Yellow Warbler, but more southerly in Alaska and northerly in Quebec and Labrador. Winters in Central America, Caribbean islands and in northern South America.

Wilson's Warbler *Wilsonia pusilla*
Breeding males have a black skull-cap, but are otherwise have yellow underparts and olive upperparts and tail (Photo 209e). Breeding females are as males, but lack the black skull-cap.

Breeding range as Yellow Warbler, but tends to be seen further north in Alaska, Quebec and Labrador. In winter the birds fly to Central America and northern South America.

Blackpoll Warbler *Dendroica striata*
Breeding males have black crowns, the rest of the head white with black streaks except on the cheeks. The underparts are white, streaked black, except on the vent. The upperparts are grey with black streaks and white barring (Photo 209f). Breeding females are as males, but lack the black crown and are duller overall.

Breeding range as Yellow Warbler. In winter the birds fly to Central and South America, some reaching southern Chile and Argentina.

Orange-crowned Warbler *Vermivora celata*
Breeding adults have the orange crown of the name, but this is often difficult to discern against the olive-green head colour. The underparts are yellow, streaked with olive, the upperparts and wings olive-yellow (Photo 209g). The overall colour varies across the range, some eastern birds being overall grey, some Alaskan birds dull brownish-olive.

Breeding range as Yellow Warbler. In winter the birds are seen in the southern United States, in Central America and in northern South America.

Buntings and New World Sparrows

'Bunting' is an old English word of debated origin and would have been taken across the Atlantic by early settlers. This resulted in some species being labelled inappropriately as buntings, and a few of these names have stuck (e.g. the birds of the genus Passerina of the southern US states, and the Blue Bunting (*Cyanocompsa parellina*). The situation was further confused by the settlers referring to some species that looked similar to birds with which they were familiar by their Old World name, in the way that, as noted earlier, the American Robin was named after its superficial similarity to the Old World Robin. In similar fashion many birds were also named 'sparrows' because they bore a resemblance to Old World Sparrows. In fact the New World Sparrows are members of the Emberizidae family which also includes the true buntings.

Emberizids evolved in the Nearctic, where most species still reside. They spread to Asia by way of the Bering Straits, and then continued as far as Europe. The birds have the stout bills characteristic of seed eaters. Males sing to establish territories, but usually have an additional repertoire of visual displays. The pair-bond is generally monogamous and seasonal, with the notable exception of Smith's Longspur. The 3–5 eggs are incubated by the female. Chicks are usually brooded only by the female, though are fed by both birds.

Here we begin with Emberizids which occur solely in the Palearctic, then consider those species which can be seen in both Palearctic and Nearctic, before dealing with the New World Sparrows.

Northern Reed Bunting *Emberiza schoeniclus*

Breeding males have black heads with a white collar and white patch below the bill base. The throat is black and there is a black bib which continues to the upper breast. The rest of the underparts are white with fine brown flecking on the flanks. The upperparts are chestnut, streaked with black and buff. Breeding females lack the head pattern, having a brown crown, finely streaked black, grey-brown supercilium to the nape, brown cheeks and a dark grey malar stripe ('moustache'). There is a white patch below the bill base, but the throat is grey. The underparts are more heavily streaked. The upperparts are buff and chestnut, with much less black. In winter males lose the striking black head pattern, looking very similar to females. The description is for the nominate species which breeds in northern Eurasia to the Ural. Eastwards are two subspecies (*E. s. passerina* and *E. s. parvirostris*), which differ largely in bill size and have a paler base colour, and *E. s. pyrrhulina* of Kamchatka which has a richer brown base colour (Photo 211a). The female of this subspecies has a rufous crown, rich buff underparts and less streaking.

An Arctic-fringe species. Northern Reed Buntings breed in northern Fennoscandia and across Russia to the Lena River (but not to the north coast), and in Kamchatka. In winter the birds fly to south-east Europe and to southern Asia, including China and Japan.

Pallas's Reed Bunting *Emberiza pallasi*

Breeding males are smaller and slimmer than Northern Reed Buntings, but similarly patterned, with the same head colour, but with a white rump and largely unstreak underparts (Photo 211c). Females have much more subdued heads, the supercilium much distinct, though there is still a malar stripe (Photos 211d and e). In winter birds are duller.

Breeds on south-eastern Taimyr and east from there to central Chukotka, but rarely to the north coast. Absent from Kamchatka. Winters in southern Russia, Korea and China.

Little Bunting *Emberiza pusilla*

Breeding adults have black crown and nape with a distinctive central chestnut streak, the rest of the head and throat chestnut, with some black streaking. The underparts are white with dark brown streaks. The upperparts are brown or grey-brown with dark brownish black streaking (Photo 211b). Summer and winter plumages are similar.

An Arctic-fringe species; Little Bunting breeds in northern Finland and across Russia to the borders of Chukotka, but not to the north coast. In winter the birds are seen in a band from northern India to eastern China.

Rustic Bunting *Emberiza rustica*
Breeding males have dark brown crown and ear coverts, but white supercilium, cheeks and throat. The lower mandible is pinkish grey, the upper mandible grey. The breast is rufous, the other underparts cream, the flanks spotted rufous. The upperparts are rufous, offset by white feather edges. Breeding females have a similar head pattern, but buff and brown replacing dark brown and white. In winter male head colour is similar to female, but darker.

An Arctic-fringe species, breeding to the treeline across Eurasia. Birds of far eastern Chukotka and Kamchatka are subspecies (*E. s. latifascia* – Photo 213a male) which has a darker crown. In winter seen in Japan and eastern China, and infrequently in Alaska.

Lapland Bunting (Lapland Longspur) *Calcarius lapponicus*
The Nearctic name refers to the long claw of the hind toe. Breeding males have black heads with a buff–white supercilium extending to the side of the nape. From the nape side to the breast there is a pale buff or white stripe. The black of the head continues over the throat and breast, though the breast sides are white. The rest of the underparts are white. The nape is bright chestnut. The upperparts are pale chestnut, streaked black. Breeding females lack the distinctive head colour, having chestnut, white and dark-brown heads. The throat and breast lack the black of the male, though there is occasionally a dark, ragged breast band. In winter males lose the striking black head pattern, looking very similar to females.

Breeds throughout Greenland (though rare in the north-east), in northern Scandinavia and across Russia, including the Arctic islands (apart from the northern island of Novaya Zemlya, but unconfirmed to date on Severnaya Zemlya). Breeds throughout Alaska, on the Bering Sea islands, and across Canada, including Arctic islands north to Ellesmere. Wintering birds are seen in a band across eastern and central Europe, southern Russia, and in the southern United States.

Snow Bunting *Plectrophenax nivalis*
The world's most northerly breeding passerine (northern Greenland and Ellesmere). A Snow Bunting was seen at the North Pole, from the deck of a surfaced US Navy submarine.

Breeding males have a white head and underparts. The upperparts are black or white heavily splattered black. The tail and wings are black-and-white. Breeding females are as males, but black is replaced with brown, and the head has spots and patches of dark brown. In winter males lose the black and white pattern and look very similar to females.

Breeds on Greenland, Iceland, Jan Mayen, Bear Island and Svalbard, in northern Scandinavia and across Russia to Kamchatka, including all Arctic islands. Breeds on the Commander, Bering Sea and Aleutian islands, throughout Alaska and northern Canada including all the Arctic islands. In winter Icelandic and some Bering Sea birds are resident, but most birds fly south to occupy broad bands across central Europe and southern Russia, and across the northern United States and southern Canada.

McKay's Bunting *Plectrophenax hyperboreus*
Once thought to be a pale subspecies of the Snow Bunting, but now considered a species. Very rare, though the population is now thought to be increasing. Much sought after by birdwatchers, but the breeding grounds are small, uninhabited islands that are difficult to reach. Named after Charles McKay, a US Signal Corps observer and bird collector who drowned in suspicious circumstances in 1883.

Breeding males are white apart from some black on the wings. Breeding females are as males but with some rufous speckling on the crown and forehead, and black streaking on the back. In winter males lose the black pattern and look very similar to females.

Rare and local. Breeds on Hall and St. Matthew Islands and, rarely, on the Pribilof Islands. In winter the birds are seen on the west coast of Alaska, from Nome to Cold Bay.

b Nominate male Lapland Bunting, northern Scandinavia.
c Juvenile Lapland Bunting of race C. l. subcalcaratus, *west Greenland.*
d Nominate female Lapland Bunting, northern Scandinavia.
e Female Snow Bunting of race P. n. insulae, *Iceland.*
f Nominate male Snow Bunting, northern Scandinavia.
g Nominate juvenile SnowBunting, west Greenland.
h McKay's Buntings in winter plumage, Bering Land Bridge Visitor Centre, Nome, Alaska.

Smith's Longspur *Calcarius pictus*

Very attractive birds, the breeding males having black-and-white head, the nape, throat and underparts of rich buff-ochre. The upperparts are dark chestnut, streaked black and buff. Breeding females have the same head pattern but in shades of brown. The underparts are less rich, and finely streaked darker. The upperparts are brighter. In winter males lose the striking black head pattern, looking very similar to females.

Although male Smith's Longspurs sing from an elevated perch as other buntings do, they do not steadfastly control a territory, with the territories of several males overlapping and even being only loosely defined. The reason is that no pair-bond is formed. When a female bird is ready to lay she will mate repeatedly (up to 50 times daily for about a week) with several males. The males also mate with any available female. The result is that a female's brood will be fathered by several males and she can expect help from all of them in raising the young.

Smith's Longspurs breed in central and northern Alaska (though rare in the north), in northern Yukon and NWT, and in a narrow band across Canada as far as southern James Bay, but no further east. In winter, seen in a narrow area of the southern USA around Texas.

Savannah Sparrow *Passerculus sandwichensis*

Breeding adults have a brown and white head with a distinct supercilium which is yellow forward of the eye, becoming white. The upperparts are brown, streaked darker, the underparts white or cream streaked dark brown. Summer and winter plumages are similar.

A true Arctic species, breeding throughout Alaska and northern Canada, and on southern Canadian Arctic islands. In winter northern birds move to the southern United States, to Central America and Caribbean islands.

American Tree Sparrow *Spizella arborea*

Breeding adults have a rufous cap and line from behind the eye, the rest of the head grey or grey-buff. The throat and underparts are grey, with rufous smudging on the flanks and a dark brown smudge at the centre of the breast. The upperparts are chestnut with darker streaking. Summer and winter plumages are similar.

Breeds in west and central Alaska (rare in the north), throughout the Yukon and much of mainland NWT, on the shores of Hudson Bay and in northern Quebec and Labrador. In winter the birds are seen in the north and central continental USA.

Harris's Sparrow *Zonotrichia querula*

The largest Nearctic sparrow. Breeding adults are handsome birds with black crown, ear patch and throat, the rest of the head grey, the bill a contrasting pink. Upperparts are chestnut and grey, streaked dark brown, underparts grey, the upper breast with a ragged black patch. Summer and winter plumages are similar.

Breeds in a broad band across Canada from the Mackenzie delta to south-western Hudson Bay. Rare in Alaska. In winter the birds are seen in the south-central USA.

Lincoln's Sparrow *Melospiza lincolnii*

Breeding adults have a rufous and grey head. The upperparts are buff and rufous, streaked dark brown. The breast and flanks are buff streaked with dark brown, the rest of the underparts white, either unmarked or with much finer streaking. Summer and winter plumages are similar.

Breeds in southern and central Alaska (but uncommon in the west), in central Yukon and NWT, around the southern shores of Hudson Bay and in central Quebec and Labrador. In winter the birds fly to the southern States, Mexico and Central America.

a Savannah Sparrow.
b Male Smith's Longspur.
c American Tree Sparrow.
d Lincoln's Sparrow. The shot was taken in the southern USA during the winter.
e Harris's Sparrow.

White-crowned Sparrow *Zonotrichia leucophrys*

Breeding adults have a distinctive crown of black-and-white stripes, the rest of the head grey. The underparts are unmarked: they are grey apart from brown flanks. The upperparts are chestnut and grey, streaked dark brown. Summer and winter plumages are similar.

White-crowned Sparrow breeds in Alaska (but uncommon in north and south) and across Canada, only to the north coast. In winter seen in the southern USA and Mexico.

Golden-crowned Sparrow *Zonotrichia atricapilla*

Breeding adults are patterned as White-crowned Sparrows, but the crown is striped yellow and black. The vent is white. In winter the striped crown is much more subdued.

The birds have one of the most distinctive songs of all Nearctic sparrows, a song which, once heard, is never forgotten and allows the observer to find the elusive singer. The song is a descending three syllables, a mournful trio, particularly when heard on a bad weather day. It is usually written as 'Oh Dear Me', which fits well with the idea that the bird is not entirely happy with its lot on such days, but could as easily be rendered as 'Three Blind Mice'.

Golden-crowned Sparrows breed in the Aleutians (to Unimak Island) and west Alaska, but are uncommon in the north and central State. Also breeds in southern Yukon and British Columbia. In winter the birds are seen along the west coast of the United States.

Fox Sparrow *Passerella iliaca*

Northern breeding adults have rufous and grey heads and upperparts, and white, rufous-streaked underparts. The bird as described is often called the Red (or Taiga) Fox Sparrow and some authorities believe it to be a species rather than a subspecies of a bird widely distributed in North America. Another species or subspecies (*P. unalaschensis* or *P. i. unalaschensis*), the Sooty Fox Sparrow, is much darker, with uniformly brown (varying from mid- to dark brown) upperparts and white underparts with heavy brown (again mid- to dark brown) streaks and spots. Summer and winter plumages are similar.

Red Fox Sparrows breed in central and western Alaska (but more rarely in the north), the Yukon and NWT, then more southerly to the southern shores of Hudson Bay and in central Quebec and Labrador. Sooty Fox Sparrows breed in the Aleutians and southern Alaska. In winter the birds fly to east and west coasts of the southern United States.

Dark-eyed Junco *Junco hyemalis*

A variable bird with significant plumage differences across the range. Males of northern (nominate) birds (often referred to as the Slate-coloured Junco) have dark slate-grey heads and upperparts. The bill is pink, the outer tail feathers white. The breast is dark slate-grey, the rest of the underparts white, smudged dark slate-grey on the lower flanks. Breeding females are brown, sometimes streaked darker on the upperparts and crown, the rest of the head and breast light slate-grey. Summer and winter plumages are similar.

Breeds in south and central Alaska (but rare in the west and the Aleutians), in Yukon to the Mackenzie delta, then southerly to the southern shores of Hudson Bay and in central Quebec and Labrador. Visitors to the Denali Park Visitor Centre will have no problem seeing the birds: indeed, it is more difficult to avoid stepping on them as they seek out crumbs dropped from snacks. In winter the birds move to the southern United States.

Song Sparrow *Melospiza melodia*

Breeding adults have grey upperparts, and white underparts heavily streaked with grey and brown so that the flanks look entirely grey and brown. As described is the subspecies *M. m. sanaka* from the eastern Aleutians, one of 24 subspecies of Song Sparrow which includes nine in the Alaska/Pacific North-West, all of which are greyer and larger than the nominate. Summer and winter plumages are similar.

Not a true Arctic breeder, Song Sparrows being at the northern extreme of their range in the Aleutian Islands/southern Alaska where they are resident.

a White-crowned Sparrow.
b Golden-crowned Sparrow.
c Red Fox Sparrow.
d Dark-eyed (Slate-coloured) Junco.
e Song Sparrow of race M. m. sanaka, *Unalaska.*

Finches and Icterids

Finches

Finches evolved in the Palearctic, the similarities of the finches to the buntings, which evolved in the Nearctic, arising in response to the evolutionary pressures of seed-eating. As some buntings have crossed to the Palearctic, so some finches have crossed the other way, with several Arctic breeders being circumpolar. Finches have an array of bill shapes, ranging from the massive conical bill of the Hawfinch (*Coccothraustes coccothraustes*), evolved for cracking large, hard seeds, to the more delicate bill of the Siskin (*Carduelis spinus*), evolved for taking grass seeds. There are also the curious bills of the Pine Grosbeak, designed for extracting conifer seeds, and the even more extreme conifer-seed extractor of the crossbills.

Finches sing and have visual displays to establish territories. Pair bonds are monogamous and seasonal. 3–6 eggs are incubated by the female only, but the chicks are usually brooded and fed by both birds. Summer and winter plumages are similar in the species described here.

Brambling *Fringilla montifringilla*
In northern Norway, where towns appear close to areas of shrub-tundra, spring visitors, after an initial surprise, become accustomed to seeing finches such as the Bullfinch (*Pyrrhula pyrrhula*) which cannot be described as an Arctic species. Only the Brambling, which breeds across Eurasia, including Kamchatka, and is occasionally seen in Alaska, to the treeline, can be considered a true Arctic (if Arctic-fringe) species (Photo 219a, female, 219b, male).

Common Redpoll *Carduelis flammea*
Breeding males have a red fore-crown, the rear crown, rest of the head and nape being tawny brown, streaked darker. The breast and upper belly are rose-pink, flecked buff, the rest of the underparts cream or buff, with dark brown streaking on the flanks and undertail. The upperbody is tawny, streaked with dark brown. Breeding female as breeding male but with no, or very little, rose-pink on the underparts. Several subspecies, differentiated by size and subtle colour changes. For breeding range see overleaf.

Arctic Redpoll (Hoary Redpoll) *Carduelis hornemanni*
Breeding adults are very similar to Common Redpoll, but paler overall, the upperparts less brown, the rose-pink of the underparts much paler, the underparts unmarked or minimally streaked. Female is as female Common Redpoll, but paler. Nominate birds breed in Greenland and eastern Canada. They are larger and usually darker than *C. h. exilipes* which breeds across remaining circumpolar range. For breeding range see overleaf.

The information above would lead the observer to believe that there would be little problem in distinguishing Common and Arctic Redpolls, but the fact that the geographical ranges overlap extensively throughout the circumpolar ranges of both, and the colour variation between individuals can mask assumed differences means that deciding what species is front of your glassware can be tricky. So tricky in fact that some experts believe that the two are conspecific, with variation in size and coloration merely indicating the unsurprising existence of subspecies across the vast range (which extends also southward, several extralimital subspecies also having been identified). The photographs here illustrate the problem, but sadly do not necessarily aid an easy solution. Photo 219d is a Common Redpoll photographed in southern Scandinavia. Photo 219c is an Arctic Redpoll (*C. h. exilipes*) photographed in northern Norway. Photo 219e is another *C. h. exilipes*, photographed in north-west Alaska. So far so good, but what about Photo 219f, which has the plumage of a Common Redpoll, but a yellow and red forehead which was taken in northern Norway and Photo 219i which seems to be an Arctic Redpoll, but was taken at the same time and in the same place as 'f' and which locals claimed was definitely a Common? There is no concrete evidence of hybridisation, but seeing these images and, most particularly Photos 219g and 219h which were both taken on the southern Taimyr Peninsula does make one wonder.

Common Redpolls breed on Greenland (but are rare in the north-east) and Iceland, in northern Scandinavia and across Russia to Kamchatka, but to the north coast only at the White Sea and in Chukotka. Breeds throughout Alaska (uncommon in the north) and across the north Canadian mainland, but only on western Baffin Island. Some Greenlandic and most Icelandic birds are resident. Other Greenlandic birds move to Iceland or join European birds in northern and central Europe. Asian birds move to Japan and south-east China, North American birds to southern Canada and the northern United States.

Arctic Redpolls breed on west and north-east Greenland (but is scarce everywhere), in northern Scandinavia and across Russia, but to the north coast only at the White Sea and in Chukotka (though also breeds on Novaya Zemlya's southern island, and on Wrangel?). Breeds in northern Alaska and across northern Canada to Hudson Bay, around Ungava Bay and Baffin and Ellesmere islands. Resident or partial migrant, moving short distances only, to southern Scandinavia, subarctic Russia and southern Canada. Some Arctic Redpolls are resident through the High Arctic winter, withstanding temperatures down to -60°C. To aid survival the birds have a storage area, called the oesophageal diverticulum, about halfway down the throat, where extra food can be stored. This allows the bird to take in more food than it can digest immediately, storing the rest for later consumption. During the long hours of night the birds fluff their feathers to increase insulation, reduce energy output by resting, and transfer the food from storage to the main digestive tract.

Crossbills have extraordinary bills and remarkably the birds do not hatch with crossed mandibles and the cross is handed, i.e. that there are left- and right- 'handed' crossbills.

Crossbills do not hatch with crossed bills as it is not the internal structure of the bill that causes the cross but the external horny layer. This develops from about four weeks of age, and is not fully developed until the chicks are 6–7 weeks old. Until that time the chicks cannot feed efficiently and must be fed by their parents. The bill has evolved to allow the birds to prise open conifer cone scales to reach the enclosed seeds allowing crossbills to feast on seeds that other species cannot reach at that stage. The bill is inserted into the cone with the mandibles open. The bill then closes, the crossed mandibles meaning the bill widens as it closes, prising the scales apart. The bird's tongue has a cartilaginous 'blade' at its tip which slices off the seed allowing retrieval. It is the lower mandible which twists in the first weeks of chick life, and the twist can be either to the left or right of the upper mandible so each bird can be left- or right-mandibled. Interestingly, if a cone is snipped off by the bird for easier manipulation, left-handed birds, having approached the cone from that side, will hold it in their left foot, and vice versa for right-mandibled birds.

Common Crossbill (Red Crossbill) *Loxia curvirostra*
Breeding males have orange-red heads and upperbody, flecked with grey. The breast and belly are similar, merging to a grey vent (Photos 221a, b and c). Breeding females are patterned as males, but the head and back are pale olive, heavily flecked grey. The rump is green, as are the underparts. The birds feed on spruce seeds. An Arctic-fringe species, breeding to the timberline in Fennoscandia and across Russia to the Sea of Okhotsk. Breeds in southern Alaska and across Canada, but rarely to the timberline. Resident and dispersive, and also irruptive.

Two-barred Crossbill (White-winged Crossbill) *Loxia leucoptera*
Breeding adults are as Common Crossbills, apart from having two distinct white bands on the inner wing. Males are also in general pinkish-red rather than red-orange (Photo 221f). Photo 221e is a female Two-barred Crossbill, but females of the other two species are very similar. The birds feed on larch seeds. Breeds to the timberline in eastern Finland and across Russia to the Sea of Okhotsk. Breeds to the timberline in Alaska and across Canada, north to the Mackenzie delta and on the southern shores of Hudson Bay. Resident and dispersive, and also irruptive.

Parrot Crossbill *Loxia pytyopsittacus*
Breeding male (Photo 221d) is as Common Crossbill, but more orange. Females are as Common Crossbill. In each case the birds are plumper, and the heads and bills are larger. Breeds to the timberline in Fennoscandia and in western European Russia. Resident and dispersive, and also irruptive.

Pine Grosbeak *Pinicola enucleator*

Breeding males have red heads with a pale grey crescent below the eye, and an indistinct dark brown crescent around the ear-coverts. The underparts are red, becoming grey on the lower belly. The upperparts are red-brown, with dark brown chevrons (Photo 223a). Breeding female patterned as male, but the head colour varies from russet to golden-yellow, the underparts and upperparts golden-yellow scalloped grey (Photo 223b).

Pine Grosbeak breeds in northern Scandinavia and across Russia to the timberline. Breeds throughout Alaska to the timberline (but uncommon everywhere and rare in the north), and to the timberline across Canada, including the southern shores of Hudson Bay and in central Quebec and Labrador. Resident or partially migratory, moving to central Scandinavia, southern Russia, southern Canada and the northern United States.

Grey-crowned Rosy-finch *Leucosticte tephrocotis*

Breeding males have a black forehead, the rest of the head silver-grey. The throat is dark brown, the underparts rufous brown, becoming redder towards the lower belly. The mantle and back are tawny brown, streaked black (Photo 223c, Unalaska). Breeding female is as male but duller overall. There are several subspecies in North America, all following the same basic plumage pattern, but with minor differences. The birds of the Pribilof Islands (*L. t. umbrina*, the Pribilof Rosy-finch) are darker and larger than other races (Photo 223d).

Breeds in western and central Alaska and throughout the Yukon, and on the Aleutian and Pribilof Islands. In winter the birds head down the west coast of the United States.

Asian Rosy-finch *Leucosticte arctoa*

Breeding adults are as the Grey-crowned Rosy-finch, but the head is overall dark grey, lacking the dark forehead, the nape is pale grey with a rosy tinge and the underparts are washed silver and pink (Photo 223e).

Asian Rosy-finches breed from the Lena to the north coast of the Sea of Okhotsk, throughout Kamchatka and on the Kuril islands. In winter the birds are seen in southern Siberia, in northern Mongolia and China, and in Japan.

Until the early 1980s the various birds of the Bering Sea were considered to be subspecies of one species, the Arctic Rosy-finch. The Russian and American birds were then split into two species, though that decision (and therefore the taxonomy) remains controversial. One specific problem relates to the birds of the Commander Islands which are very large (the size of Starlings – Photo 223f). Are they a subspecies of the Grey-crowned Rosy-finch or of the Asian Rosy-finch? And if they are a subspecies of the latter, where does that leave the Kamchatka birds?

Icterids

The icterids are a family of more than 100 Nearctic species that includes some of the most abundant birds on Earth [e.g. the Red-winged Blackbird (*Aegalaius phoeniceus*) which has an estimated population of around 200 million]. One species breeds at the Arctic fringe.

Rusty Blackbird *Euphagus carolinus*

Breeding male is black, the head with a purple sheen. Breeding female brown, darker on the upperparts than the underparts, the latter occasionally with fine barring (Photo 223g). In winter males are overall dark brown, often with a discernible rusty wash that gives the bird its name. In winter females are paler than in summer.

Rusty Blackbirds breed throughout Alaska, but uncommon everywhere and very rare in the north. Also breeds in Yukon to the Mackenzie delta, then southerly to the southern shores of Hudson Bay, in central Quebec and northern Labrador. In winter the birds fly to the south-east United States.

Shrews

Shrews are the only members of the Insectivora that breed in the Arctic. Apart from the Water Shrew, Arctic shrews all belong to the genus Sorex, the red-toothed shrews, so called because the tips of their teeth are red, a feature caused by iron compounds in the enamel.

Shrews are small and mouse-like, but with long, mobile noses. Their sense of smell is acute, but their eyesight is poor. Some species use ultrasonic echolocation for both navigation and in the search for prey, though it is not clear how sophisticated these systems are as the shrews are one of the least-studied mammal families, almost all that is known about them deriving from studies of just a handful of species. Some shrews produce a neurotoxin from their salivary glands. This is injected into the prey when the shrew bites; immobilisation of prey is important, as shrews often take relatively large prey. Species with the most toxic poisons take small vertebrates as well as the invertebrates that make up the 'standard' shrew diet.

Shrews have a cloaca, a single opening for the urinary, digestive and reproductive tracts (as do most birds). Shrews also practice refection, the re-ingestion of excreted food. To do this, the shrew curls into a ball and licks the cloaca until the rectum is extruded. The rectal contents are then re-ingested. On a second pass through the digestive tract the re-ingested matter is digested further before being excreted as faeces. Refection only occurs when the digestive tract is free of faeces.

Shrews have a very high metabolic rate (and a heart rate which may reach 1000 beats/min.) and must feed very frequently. Indeed, some shrews will actually die of hunger if they are unable to feed every one or two hours, the shrews life being foraging interrupted by short intervals of rest. Hunting is haphazard, a near-constant scurrying, the shrew taking anything that crosses its path which can be immobilised and consumed. The iron-rich red-tipped teeth of the Sorex shrews are a necessary adaptation for the genus. In general these species have a higher metabolic rate than other shrews and as a shrew does not replace its teeth (the enforced starvation of tooth-loss would result in death), the harder tooth-tips of Sorex shrews reflects the higher food intake required. If the shrew lives long enough it will wear its teeth down and die of starvation. Shrews cannot hibernate, since their metabolic rate cannot be lowered sufficiently to allow survival on fat stores. The Arctic species shrink both their skeletons and internal organs to reduce food requirements, but even so many die of starvation during the winter.

Despite the extreme metabolism of shrews, which would seem to preclude life in as harsh an environment as the Arctic, there are species which are true Arctic dwellers. One is the American Masked (or Cinereus) Shrew (*Sorex cinereus*: Photo 225c) which breeds in Alaska to the west coast, north to the Brooks Range, on the Aleutian Islands, near the Mackenzie delta, then follows the timberline across Canada. Shrews are elusive and studying them in the field is difficult, but it is now believed that a shrew once bred across Beringia, and that the shrews of St. Lawrence Island (*S. jacksoni*), the Pribilofs (*S. hydrodamus*), the Barren-ground Shrew (*S. ugyunak*) – which breeds along the North American Arctic coast to the Boothia Peninsula: ugyunak is shrew in Inuktituk, and it is the only species the Inuit were likely to have seen – Portenko's Shrew (*S. portenkoi*) of Chukotka and the Kamchatka Shrew (*S. camtchaticus*), are all likely to have derived from the ancestral *cinereus*. Similarly in Eurasia, the Eurasian Masked (or Laxmann's) Shrew (*S. caecutians*: Photo 225a) is likely to be a close relative of the Flat-skulled Shrew (*S. robaratus*) of upland eastern Siberia. The Tundra Shrew (*S. tundrensis*: Photo 225d) of the east Siberian tundra (a true Arctic dweller), is likely to be related to the Common Shrew (*S. araneus*) of western Eurasia and to the (somewhat misnamed) Arctic Shrew (*S. arcticus*: Photo 225b) of North America which breeds to Hudson Bay's southern shore.

In addition to the above, several other shrews may be seen as far north as the timberline: the Taiga Shrew (*S. isodon*) and Large-toothed Shrew (*S. daphaenadon*) of Siberia and Kamchatka; the American Pygmy Shrew (*S. hoyi*) which breeds across the continent; the Pygmy Shrew (*S. minitus*) of northern Fennoscandia, but more southerly across Russia to Lake Baikal; the Dusky (or Montane) Shrew (*S. monticolus*) of south-west and central Alaska, south and (less commonly) northern Yukon and southern NWT; and the Water Shrew (*Neomys fodiens*) of Northern Fennoscandia and southern Russia.

Rodents

With a species list that comprises more than 40% of all mammals, the rodents are one of the most successful and widespread of all the mammalian orders. They are also one of the most numerous, a consequence of their prodigious ability to breed. Female rodents can often breed when only six weeks old, have a gestation time of only c.20 days, and are often sexually receptive within days of giving birth. Studies on mice have shown that a single breeding pair can yield a population in excess of 500 within six months. Rodents vary greatly in size (the smallest weigh just a few grams, while the Capybara (*Hydrochoreus hydrochaeris*) can weigh 90kg) and form, but they do have features in common, one of which is that in both the upper and lower jaws there is a pair of incisors that grow continuously, requiring the animal to gnaw in order to maintain incisor length – the word rodent derives from the Latin *rodere*, to gnaw. If the animal were to stop gnawing, the teeth of the lower jaw would eventually grow into the brain, while those of the upper jaw would grow through the lower lip, though in either case death through starvation would happen first. Feeding generally provides all the gnawing needed, but rodents will sometimes gnaw other materials to help keep their incisors a constant length. Rodents have an essential ecological role. As examples, they are pivotal in supporting many populations of predatory mammals and birds, and several mycorrhizal fungi (the most famous of which are the truffles) are spread only by being consumed and then excreted by rodents. As the fungi are necessary for the growth of plants, the rodents play an important role in the spread and establishment of plants; the recolonisation of land uncovered by the retreating ice of the last Ice Age was hastened by the spread of mycorrhizal fungi.

Rodents are classed into suborders defined by the bone/muscle systems which control their gnawing jaws. The suborders can be generalised as 'squirrel-like', 'cavy-like' and 'mouse-like'. The first of these include several Arctic and Arctic-fringe species – squirrels, ground squirrels, marmots and beavers. The second only one Arctic-fringe species, the North American Porcupine, while the third includes a host of Arctic-dwelling lemmings and voles.

Squirrels and Marmots

Eurasian Red Squirrel *Sciuris vulgaris*
Adults are red-brown overall, but paler on the underparts (Photos 227a and b). In winter the pelage is darker. The tail is bushy. There are usually ear tufts, these being more prominent in winter. Breeds to the treeline in Scandinavia and across Russia to Kamchatka.

American Red Squirrel (Spruce Squirrel) *Tamiasciuris hudsonicus*
Adults are red-brown above (usually darker and less 'warm' than the Eurasian Red Squirrel), paler below (Photos 227c and d). There is sometimes a black band separating the upper and lower parts. In winter the pelage is brighter. The tail is bushy. In winter the ear tufts are more prominent. Breeds to the timberline across North America.

Black-capped Marmot *Marmota camtschatica*
The only marmot in the northern Palearctic. Adults have brown upperparts, with a distinctive black or dark brown cap. The cap actually covers both crown and nape, and extends below the eye (Photo 227e).

Rare, with a curiously patchy distribution that includes the mountains and tundra of Baikalia, the Upper Yana and Kolyma rivers, northern Kamchatka and parts of central and northern Chukotka. Essentially an upland animal, found at altitudes to 1,900m. Black-capped Marmots excavate extensive burrow systems that may be up to 100m long and go into the permafrost. See comments overleaf regarding hibernation.

Hoary Marmot *Marmota caligata*
Adults may be brown or grey-brown overall, but with darker faces and, usually, a red-brown tail (Photos 227f and g). Breeds in southern Alaska and Yukon. Often seen in the Denali National Park.

Alaska Marmot (Brooks Range Marmot) *Marmota broweri*

Smaller than the Hoary Marmot. Adults are grey with silvery guard hairs, which give a frosted appearance. The head, back, rump and hind legs are often darker than the upper body. The head has a distinctive black cap and nose.

Marmots are the largest of all true hibernators. For Alaska Marmots hibernation begins in September, with the animals sometimes remaining in their burrows until early June, having spent almost nine months underground. The Black-capped Marmots of eastern Eurasia may spend even longer in their burrows: indeed, so long is the underground period of this species (altitude adds to latitude to create an environment that is harsh almost throughout the year) that they breed and may even given birth before they emerge for the brief summer. For all marmots except the Woodchuck (*Marmota monax*), whose range is the most southerly of the group, the summer is so short that the young do not mature sufficiently to breed at one year and begin breeding only when they are two. For the Black-capped Marmot the short summer does not allow sufficient resources for females to breed annually, so instead breeding every two years is the norm.

Alaska Marmots breed in the Brooks Range, Alaska, particularly on the northern slopes where the boulder fields and talus slopes offer the perfect habitat, the stony ground making digging difficult for Brown Bears which would otherwise excavate the burrows. There are unconfirmed reports of breeding to the northern coast. If such breeding is confirmed, the species would be the most northerly of all marmots, exceeding the furthest north confirmed breeding of Black-capped Marmots. The marmots are also seen on lower slopes/foothills of the range, Photo 229a having been taken in late May near Kotzebue when the animals can become tolerant of people (but not dogs) if unharassed. It has been conjectured that some marmots may move to burrows at lower elevations to take advantage of earlier spring vegetation.

Arctic Ground Squirrel *Urocitellus parryi*

Note: the scientific name of the species is one of several synonyms that may be seen in the literature, the most frequent alternatives being *Spermophilus parryi* and *Citellus parryi*. That given here is the one recently agreed as most appropriate.

Adults are tawny brown with buff and white flecking. The head is darker, the forehead and face usually red-brown. The tail is short (the occasional name Long-tailed Souslik deriving from the fact that, though relatively short, it is longer than in some other ground squirrels and sousliks). Though primarily vegetarian, the animals have been seen feeding on a Caribou carcass, and feed on dead birds and dead of their own kind, as well as on insects. Males will also kill and eat young they have not fathered. The squirrels will also visit camp sites and steal available high fat food stuffs, and may even become reasonably tame if unmolested.

A true Arctic dweller, the squirrels form colonies of up to 50 individuals living in underground burrows, with numerous entrances, and also additional 'blind' tunnels, to offer some protection against predation. Such colonies are usually under the control of a dominant male. The squirrels hibernate in the burrows in a nest of vegetation, laying down a store of seeds prior to sleeping so that they can feed as soon as they wake. Recent studies have shown that while both sexes spend more or less equal times above ground, it is the females that do the majority of the gathering of food stores and the rearing of young, the males tending to bask in the sun. The finding seems to reinforce female prejudices, but, the research team suggested, may be influenced in some way by ground squirrel society. Given that virtually all Arctic terrestrial and avian predators take Arctic Ground Squirrels, the male behaviour would otherwise appear somewhat reckless.

Arctic Ground Squirrels breed in Chukotka and Kamchatka, in northern Alaska and northern Canada to the eastern shores of Hudson Bay. Absent from Canada's Arctic islands. There are numerous subspecies across the range.

a Alaska Marmot.
b Arctic Ground Squirrel, west Alaska.
c Arctic Ground Squirrel, Yttagran Island, Chukotka.
d-g Arctic Ground Squirrels, northern Canada.

Beavers

Although the beaver is a subarctic animal – its lifestyle means it cannot thrive beyond the treeline – it is likely to be seen by Arctic travellers close to the northern limits of the forest, for instance near the Mackenzie delta, and in the Denali National Park. The most distinctive feature of the beaver is the long, flattened, scaly tail, which acts as a paddle and also as a fifth leg, allowing the animal to stand gnawing trees. Other adaptations include the closure of the beaver's ears and nose when it dives, while membranes cover the eyes. The lips close behind the incisors and the throat structure allows the tongue to act as a 'plug' so the animal can manipulate branches underwater to build its dam without ingesting water.

Beavers are social, living in families of up to 12, though 5–6 is more usual. The family holds a territory that includes not only a lake or section of river, but the ground to the sides from which the animals harvest trees, mud and rocks to construct a dam (Photo 231e, Canada). Beavers may excavate channels or divert streams within their territories, as they can move more easily in water than on land and it is also easier to transport dam material. Dams can be up to 4m high, 6m thick at the base, and up to 100m long. Trees of 1.2m diameter have been seen in dams: such huge logs are not hauled, but felled close to the dam. The dam is not watertight, so less susceptible to destruction if the river volume increases, but is sufficiently flow-resistant for water to back up behind it, forming a pond. In this, or at the edge of a lake if the beavers do not need a dam, the beavers build a conical lodge of branches and mud (Photo 231f, Sweden), with an underwater entrance to above-water living quarters. In the pond the animals cache fresh branches for eating during the winter (beavers are herbivorous: they do not fish in their ponds). Ultimately the pond behind the dam silts up. The beavers then abandon it, the site ultimately reverting to forest, or becoming meadowland.

Historically the animals were hunted for their pelts and as food – beavers are large (up to 30kg, only Capybara are larger rodents) and so provide a good return in meat for the effort of trapping – and for *castoreum*, a secretion of the preputial gland that contains salicin (the basis of aspirin) that has been used as a medicine since at least the time of Hippocrates. In North America, beaver pelts are the basis of the wealth of the Hudson's Bay Company, whose officers helped explore and map Canada and occasionally offered assistance (sometimes life-saving assistance) to British Royal Navy expeditions searching for the North-West Passage.

American Beaver   *Castor canadensi*
Adult is dark brown overall, with some animals almost black. The blunt muzzle is usually paler (Photos 231c and d, Alaska). The tail is dark grey-brown or black. Breeds to the timberline across North America. Has also been introduced into lakes in Finland.

Eurasian Beaver   *Castor fiber*
Adults are paler and more red-brown than American Beavers, though some animals may also be dark brown (Photos 231a and b, Sweden). Apart from colour the two species are indistinguishable. Breeds to the timberline in Scandinavia and Russia (extremely scarce in central and eastern Russia), but the decimation of the species – in the early years of the 20th century it is estimated that the entire population throughout the range numbered no more than 1,000 animals – means that the present distribution is very patchy.

Porcupine

Though an Arctic-fringe species Porcupines may occasionally be seen on the tundra if there is dwarf willow or dwarf alder. The animals have poor eyesight, but good hearing and sense of smell. Contrary to what is occasionally written, the animals cannot 'throw' their quills. When threatened they erect the quills, and turn away from the threat so as to swing their long, heavy and formidably quill-ladened tails in the hope of hitting the aggressor. Quills that connect may detach causing long-term problems for the predator.

North American Porcupine   *Erethizon dorsatum*
Absolutely unmistakable (Photos 231g and h). Breeds across Alaska and northern Canada, including the Arctic National Wildlife Reserve in northern Alaska.

Microtines

Lemmings and voles make up the largest fraction of the Microtinae rodent sub-family, and they include the most northerly of all rodent species. Most microtines, and all Microtus rodents, have, in addition to the incisors that characterise rodents, continuously growing molars, which aid the grinding of tough vegetation. Too small to be able to hibernate without starving, microtines are active throughout the winter, using a series of burrows beneath the snow to forage. Spring snow melt reveals these burrows (Photo 233d). The snow blanket offers excellent insulation against the Arctic winter, and also some protection against predation. However, some northern owls are able to detect rodents aurally beneath a significant layer of snow, mustelids can follow them in their burrows, and Arctic Foxes also pounce through the snow to reach them: the life of a microtine is one of constant threat from starvation and predation. Many Arctic predator populations are 'locked' into microtine populations, so that the regular sharp increases in rodent populations are mirrored by increases in predator numbers. For the Arctic traveller this can be a benefit: microtines are hard to spot, but become much more visible during 'lemming years', while the predators that are often the most sought-after species become more abundant.

There are many northern microtines, all with the same basic characteristics of being small, stocky and brown. Differentiating between them can therefore be difficult, particularly if a view of them consists only of a glimpse of a disappearing animal. In the section below, the basic details of colouring only are given, together with range – the latter is often more useful in deciding what it was that you almost just saw.

Lemmings

Norway Lemming *Lemmus lemmus*
A beautiful animal, exquisitely coloured. Golden brown above, paler below, the upperparts marked in black, with black crown and forehead, saddle and stripes on the flanks. The feet are silvery–white (Photos 233a–c). The myth that the animals commit mass suicide arises from the regular population surges (in lemming years). Norway Lemmings are very intolerant of their fellows, older lemmings forcing younger ones out of territories. The topography of Norway, with its high ridges separating narrow valleys, concentrates migrating animals. Conflicts result and the population becomes increasingly panicky. When a stream, river or lake is reached, pressure from following animals causes the leaders to start swimming. If the lake is large, or if it is the sea, the lemmings, with no concept of lake size or of an ocean, swim regardless, behaviour seen as suicidal by human observers who know that the animals will drown before reaching the far side.

Breeds in northern Fennoscandia, but essentially a subarctic species, breeding in central Norway and Sweden.

Siberian Brown Lemming *Lemmus sibiricus*
Similar to Norway Lemming, but a more subdued yellow-brown, with a thick black stripe on the head and back (Photo 233f). Breeds on the northern coast of Russia from the White Sea to the Kolyma, on Novaya Zemlya, the New Siberian Islands and Wrangel.

North American Brown Lemming *Lemmus trimucronatus*
Very similar to Siberian Brown Lemming (with which it was once considered conspecific). The pelage is tawny with buff and darker brown flecking, paler on the flanks and paler again on the underparts. Breeds in western and northern Alaska and across northern Canada to the western shore of Hudson Bay, on southern Canadian Arctic islands from Banks to Baffin, and on islands in the Bering Sea.

Wood Lemming *Myopus schisticolor*
Uniformly dark grey, with paler underparts. Develops a rusty patch on the lower back, but this is much more indistinct than in the brown lemmings (Photo 233e). Breeds in southern Scandinavia, but more northerly in Russia, though only to the north coast near the White Sea and the Lena and Kolyma deltas. Breeds throughout Kamchatka, but largely absent from Chukotka.

Northern Bog Lemming *Synaptomys borealis*
Mouse-like (apart from the tail). Grey-brown above, with occasional patches of grey and of red-brown, and grey below. The tail is bi-coloured, as in many vole species. Breeds in southern and central Alaska and across central Canada to southern Hudson Bay; more northerly on the bay's eastern shore.

Collared Lemmings

Collared lemmings, true Arctic animals, belong to the genus Dicrostonyx, the name derived from the Greek for 'forked claw', a reference to the growth of a double claw in winter to aid digging in the snow. Most species have a 'collar', but if pale this merges with the colour of the upper body and so is visible only as a chest-band.

Northern (or Greenland) Collared Lemming *Dicrostonyx groenlandicus*
Similar to Arctic Collared Lemming but greyer, with buff flanks and chest and pale grey head (Photo 235a). Winter pelage is white. Breeds in north Greenland, Aleutian and Bering Sea islands, west and north Alaska, north Canadian mainland east to Hudson Bay and all islands.

Ungava Lemming *Dicrostonyx hudsonius*
Red-brown or, more usually, grey-brown. Winter pelage is white. Breeds on the Ungava Peninsula, in Labrador, and on islands of eastern Hudson Bay.

Richardson's Collared Lemming *Dicrostonyx richardsoni*
Dark red-brown above, paler on the flanks and buff or buff-grey below. Breeds around the western shore of Hudson Bay.

Arctic Collared Lemming *Dicrostonyx torquatus*
Variable, usually red-brown on the back (but grey-brown towards the lower back and on the head) with a dark dorsal stripe from the head to the tail (Photo 235b). Winter pelage is white. Breeds from White Sea to Chukotka, in eastern Kamchatka, on the southern island of Novaya Zemlya and throughout Severnaya Zemlya and the New Siberian Islands.

Wrangel Island (or Vinogradov's) Lemming *Dicrostonyx vinogradovi*
Status: IUCN: Data deficient.
Photo 235c. Breeds only on Wrangel Island, where they are rare.

Voles

As with other rodents, differentiating voles is a job for experts. Voles differ from lemmings in having longer tails, which are invariably bi-coloured.

Singing Vole *Microtus miuris*
Variable, usually dark brown above, more tawny on flanks with grey-buff underparts (Photo 235d). Adults occasionally sit in exposed places and 'sing' a high-pitched trilling 'song'. This may be an alarm call for young voles, as the singing usually takes place when litters have been weaned. Before winter arrives, the voles make hay balls (which can be huge – up to 30 litres) to provide winter sustenance. Breeds in north and south Alaska, but absent from most of the central State, Alaska Peninsula and Aleutians. Also breeds in central Yukon.

Grey-sided (or Grey Red-backed) Vole *Myodes rufocanus*
Grey above and on flanks, with a broad chestnut band on the forehead, crown, nape and along the back to the base of the tail. The underparts and feet are pale grey. The tail is chestnut brown above, grey below (Photo 235e).
  Grey-sided Vole breeds in northern Scandinavia and across Russia to Kamchatka, but only to the north coast on the Kola Peninsula and east to Baydaratskaya Bay.

Insular Vole *Microtus abbreviatus*
Almost identical to Singing Vole, having brown upperparts, yellow-brown flanks and pale buff underparts.Breeds only on the islands of Hall and St Matthew in the Bering Sea.

Northern Red-backed (or Ruddy) Vole *Myodes rutilus*
Pale golden-brown above with a broad red-brown band from the forehead, across crown and back to the tail base (Photo 237a). Breeds in northern Scandinavia, across Russia, in Alaska and northern Canada east to Hudson Bay. Absent from Arctic islands.

Southern Red-backed Vole *Myodes gapperi*
Coloured as Northern cousin, but paler (Photo 237b). Subarctic, but breeds to the southern shores of Hudson Bay and on the eastern side of the bay in central Quebec and Labrador.

Narrow-headed Vole *Microtus gregalis*
Variable, light ochre to dark brown above with characteristic darker spotting and white tips to individual hairs, which give an overall silvery appearance. The skull is narrow, particularly between the eyes (a diagnostic characteristic). Has a patchy breeding distribution across Russia, close to White Sea coast and east to the Ob delta, from Khatanga to the Kolyma delta. In Kazakhstan and Kyrgizia mountains the species breeds to 3,500m.

North Siberian Vole *Microtus hyperboreus*
Grey-brown above (often with a reddish wash), silver-grey below, this occasionally washed yellow. Has a long, bi-coloured tail. Breeds in Russia from the Ob River to the western edge of Chukotka, but only to the north coast near the Lena and Kolyma deltas.

Tundra (or Root) Vole *Microtus oeconomus*
Dark brown above with buff flecking, paler on the flanks and pale buff below. American animals tend to be larger, but with a shorter tail (Photo 237c). Breeds in northern Scandinavia and across Russia to Kamchatka (to the north coast except on the Taimyr Peninsula), and to the north coast throughout Alaska and the Yukon.

Meadow Vole *Microtus pennsylvanicus*
Red or dark brown above and on the flanks, with pale grey-brown underparts (Photo 237d). Breeds in central Alaska, northern Yukon, then more southerly to the west coast of Hudson Bay, around the bay's southern shore and in southern Quebec and Labrador.

Lemming Vole *Alticola lemminus*
Grey-brown overall, including the tail, which ends with long bristles (Photo 237e). The winter pelage is much lighter and may even be white. Breeds in a broad semi-circular belt from the Lena, curving towards the northern shore of the Sea of Okhotsk then north into Chukotka, reaching the Bering Sea as far north as Cape Dezhnev. There is a second belt that follows the western shore of the Sea of Okhotsk, then inland towards northern Lake Baikal.

Taiga (or Yellow-cheeked) Vole) *Microtus xanthognathus*
Dark brown or grey-brown above, paler below, with prominent yellow-brown cheek-patch. Breeds in central Alaska, north Yukon and boreal belt to the south-west Hudson Bay.

Sibling (or East European) Vole *Microtus levis*
Dark brown above, paler on the flanks and grey below (Photo 237f). Breeds in southern Scandinavia and south-west Russia and is non-Arctic. However, colonies have been found on Svalbard. It is assumed that these have been accidentally introduced from ships. The voles have become established, countering high winter mortality by increased fecundity.

Muskrat *Ondatra zibethicus*
The largest microtine, and also one of the easiest to observe. The name derives from the odour of the scent glands used to define territories. Adult is a rich red-brown above, becoming paler on the flanks and much paler on the underparts (Photos 237g and h). The tail is laterally compressed (about three times as deep as it is wide) as an aid to swimming as Muskrats are primarily aquatic. Chiefly vegetarian but also take molluscs. Breeds from west Alaska to Labrador, but not to the north coast. Because the animals were valued for their fur (musquash), Muskrats were farmed in Europe and escapees have produced breeding populations in Scandinavia and across Russia.

Lagomorphs

Lagomorphs – rabbits, hares and pikas – differ from rodents in having a second pair of incisors known as peg teeth. While picas look very similar to rodents, hares and rabbits differ in body shape, having long back legs that allow them to run fast. Though the adaptation is the same in each group, the strategy for avoiding predators differs. While hares simply seek to outrun a predator, rabbits make short runs to their burrows or to available cover. These different escape strategies are reflected in the breeding biology of the two groups. Young rabbits are born underground and are altricial. By contrast, leverets are born above ground and are precocial, the latter a necessity for the Arctic species as the climate, even in summer, can be harsh. To avoid drawing unnecessary attention to the leverets, the female hare visits them infrequently, usually only once each day. At that time the leverets, which spend the rest of the day in concealed places, congregate at their place of birth and are suckled for about five minutes. Because of the limited duration of nursing, hare milk has a very high fat and protein content and the leverets are weaned at three or four weeks. Northern species differ from their southern cousins by having, in general, larger litters and shorter gestation periods as insurance against higher mortality (predation and weather both taking a toll) and the short Arctic summer.

Despite looking very similar to voles and lemmings, pikas are closely related to hares and rabbits. They are distinguished from rodents by their larger ears and by having no tail, but anyone lucky enough to observe a pika eating will also notice that their jaws move side-to-side in the same way as other lagomorphs. Pikas are diurnal and very active. Males hold territories which they defend vigorously. Both sexes are very vocal, emitting a sharp whistle which may be used in territorial disputes, to initiate or reinforce pair bonds, or as an alarm call as pikas are predated by mustelids, foxes and birds of prey. The whistle is the source of the curious name, given to animals by the Tungus people of Siberia as *peeka*, uttered quickly and high-pitched is a reasonable representation of the noise. The whistles will alert observers to the presence of pikas, but they are more often heard than seen as they move around within boulder fields and talus slopes. The daily activity is, in part, a preparation for winter as the animals collect vegetation which is formed into 'hay stacks' for use as winter forage.

As with shrews, lagomorphs practice refection, the redigestion of 'first-pass' faecal material, which is partially undigested. The 'second-pass' faecal material is in the familiar form of hard, black 'currants'.

Arctic Hare *Lepus arcticus*

In summer adults in the south of the hare's range are grey or grey-brown. In winter, they are white apart from the ear-tips and inner ear. Leverets are pale grey. Arctic Hares can reach speeds in excess of 6okm/h and can swim well. In northern parts of the range the hares are white at all times, the limited summer and poor forage of the polar desert meaning that there is neither time or nourishment available to expend energy in moulting into and then out of a winter pelage. This means that the hares are extremely visible in summer. As the High Arctic wolf is a predator of the hares, this seems unfair, but it appears nature has compensated by forcing the wolves to make the same decision: High Arctic wolves are also white at all times and so are themselves highly visible.

Arctic Hares breed on the central and northern coasts of both east and west Greenland, but are rare in the south. Also breeds on Canada's northern mainland, and the Arctic islands to northern Ellesmere. The mainland population breeds as far south as Hudson Bay's southern shore. Absent from Alaska. In winter Arctic Hares occasionally burrow into the snow, excavating short tunnels to dens that offer shelter from wind and intense cold. However, snow and ice themselves do not represent a hazard to the animals, hares being occasionally seen several kilometres out on sea ice.

a Arctic Hare, Ellesmere Island. This photo and 'e' show how easily seen the hares are in summer in the High Arctic.
b, c Arctic Hare moulting from winter to summer pelage, south-west Hudson Bay.
d Arctic Hare in winter pelage, Canadian Barren Lands.
e Arctic Hare Levert, Ellesmere Island.
f Young Arctic Hare, Victoria Island.

Alaskan Hare *Lepus othus*

The Alaskan equivalent of the Arctic Hare, though much larger, this particularly noticeable with the head, which is massive: at up to 7kg, the Alaskan is one of the largest of all hares. In summer adults are similar to Snowshoe Hares, being red-brown above, paler below, with a darker crown. The ears are bi-coloured, the front half as the body, the rear half white. The ear-tips are dark brownish black. As the ranges of Alaskan and Snowshoe overlap, the Alaskan can be distinguished by sheer size, by the fact that its tail is white or pale grey in summer and by not having the extraordinary rear feet of the Snowshoe. In winter the Alaskan is white, though the ear colour is maintained. Alaskan and Arctic hare ranges do not overlap, so there is no likelihood of confusion.

Alaskan Hare breeds in south-western, western and northern Alaska. The occasional name of Tundra Hare indicates the species' preferred habitat and also adds identification against the Snowshoe, which is essentially a forest species.

Snowshoe Hare *Lepus americanus*

In summer adults are red-brown above, with pale grey-brown underparts, including the chin. The ears are relatively small for a hare, and black-tipped. The winter pelage is white apart from the black ear-tips. As noted above, size can be a distinguishing feature in areas of range overlap with the Alaskan Hare, female Snowshoes rarely exceeding 2kg (males are significantly smaller, often only 70-80% of female size). The hind feet are large (disproportionately so – hence the name, as the hare looks as if it were wearing snowshoes), an adaptation to allow the hare to move easily over soft snow. The hare is fast, with speeds of up to 40km/h claimed and at speed can make single bounds of 3m when escaping predators. In winter Snowshoe Hares occasionally burrow into snow to escape bad weather.

Snowshoe Hare breeds across North America, but is rarely seen above the timberline. While most hares are nocturnal, or mostly so, Snowshoes are very nocturnal, sheltering in thickets or under fallen trees and emerging only around sunset. Daytime sightings are therefore rare other than at dawn or dusk.

Mountain Hare *Lepus timidus*

In summer adults are dark brown or grey-brown, paler below and with a grey or white tail. The ears are black-tipped. The winter pelage is white (apart from the ear tips), but not white to the body (as it is for the Arctic Hare), the underfur being slate-grey or blue-grey. This is occasionally visible, giving greyish blue patches on the flanks. Animals in northern Siberia are white at all times.

Mountain Hares breed in northern Scandinavia and across Russia to Kamchatka, to the northern coast, but absent from all Russia's Arctic islands.

Northern Pika *Ochotona hyperborea*

Adults are uniformly brown or dark brown, occasionally with darker patches on the sides of the neck and a darker wash to the lower back. There are white stripes around the periphery of the large ears (the ears, inconspicuous tail and very long whiskers of pikas distinguish them from voles). The winter pelage is usually grey-brown or reddish-brown, the underparts paler.

Northern Pikas breed in Russia from the Yenisey River to Chukotka and Kamchatka, north to the coast (and south into Mongolia).

Collared Pika *Ochotona collaris*

Adults are a uniform grey-brown, though usually with a reddish wash. There is usually a paler collar and the underparts are also often paler. The feet are grey.

Collared Pika is much less Arctic than its Palearctic cousin, breeding in southern and central Alaska (including the Denali National Park), and southern Yukon.

a Alaskan Hare in winter pelage.
b Mountain Hare.
c Snowshoe Hare.
d Northern Pika.
e, f Collared Pika.

Ungulates

Caribou/Reindeer

The large terrestrial herbivores of the Arctic are artiodactyls, or even-toed ungulates, hoofed animals of the families Cervidae and Bovidae (the other ungulate group, the perissodactyls or odd-toed ungulates, has no Arctic representatives). The Cervidae or deer are distinguished from other artiodactyls by the antlers that are grown by males (and females in Reindeer). In general, antlers are grown and shed annually. They are horn-like extensions that grow out from the frontal bone. The antlers grow within a sensitive, blood vessel-rich covering called velvet. Ultimately the velvet is shed, often hanging in shreds from the antlers for several days until it is finally rubbed free by the deer. In southern deer species the antlers are rubbed against trees, staining them (and damaging the tree), while northern species have to be content with rubbing against the ground. The reason for the evolution of antlers is still debated. Antlers are used during the rut, the annual mating ritual when males compete to establish and maintain harems of females, mating each time a female becomes receptive. Although many competitions involve merely the showing of the antlers to a rival, accompanied by a bellowing that indicates the fitness of the male (the lower pitched the bellow, the bigger and more worthy the male), occasionally males will lock antlers and engage in battle. However, battles are usually trials of strength, the male forced backwards accepting defeat and retreating: antlers are fearsome weapons and could inflict considerable damage, so the males are keen to avoid real conflict. That suggests an alternative reason for antlers – that they are a good guide to mating fitness: this has been confirmed by studies that show that females whose partners had impressive sets produced young that were stronger and matured faster. At the end of the rut bone-dissolving cells invade the base of the antlers, and they are then shed.

Uniquely among deer, both male and female Reindeer grow antlers. The reason appears to lie in the need to excavate holes in the snow to reach winter browse. The larger males might allow females to dig holes, then evict them for themselves. Antlers allow the female to prevent such piracy. To support this, females maintain their antlers for longer than the males, usually throughout the winter. Furthermore, females of forest-dwelling species (*R. t. caribou* and *R. t. fennicus*), which feed on lichens growing on tree trunks, often do not grow antlers – there is always another tree to browse. As with other deer, the energetic cost of producing antlers is considerable, and Reindeer often eat their shed antlers to regain the minerals within. (As an aside, since female Reindeer not only possess antlers but maintain them through the winter, Santa's Reindeer is much more likely to be Rachel than Rudolf.)→

Reindeer (Caribou) *Rangifer tarandus*
Status: IUCN: Vulnerable (population declines of 40% (Alaska), 50% (Canada), 30% (Greenland), 20% (Russia), chiefly in forest species due to increases in forestry and industrialisation. Greenland and all tundra animals are vulnerable to climate change. Scandinavian and Russian subspecies are vulnerable as they are remnant wild populations).

The Reindeer of the Palearctic and the Caribou of the Nearctic are the same species. The North American name derives from *xalibu*, which means 'pawer' in Mi'kmaq (Nova Scotian native American), a reference to the animal's pawing search for winter browse. Adults are dark brown or grey-brown, but paler on the underparts, rump and feet and usually darker on the muzzle and forehead. The tail is surprisingly short, leaving the deer defenceless against mosquitoes and parasitic flies. In winter Reindeer moult to a paler pelage.

a, b and d Barren-ground Caribou (R. t. groenlandicus), in winter and spring, north Canadian mainland.
c Caribou on Bylot Island, Canada. While probably a hybrid of Barren-ground and Peary Caribou, the size, pale colour and solitary nature of the animals on the island is suggestive of pure-form Peary. Note the similarities to Photos 245d and e.
e Grant's Caribou (R. t. granti), north-western Alaska.

Reindeer have the widest feet of any deer and have well-developed dew claws (which are vestigial in most deer species): these features are adaptations for walking on snow, the hooves are also sharp-edged to aid walking on ice. As the animal walks, a tendon stretches across a bony nodule, causing a characteristic click.

There are several Reindeer subspecies. Tundra (nominate) animals are the best known, but in both North America and Eurasia there are Reindeer that forage in forests – the Woodland Caribou (*R. t. caribou*) and Forest Reindeer (*R. t. fennicus*). Recent interviews with the Dene folk of NWT also suggested a third subspecies - Mountain Caribou: genetic analysis has proved the Dene to be correct. Three subspecies also breed on islands, one on Svalbard (*R. t. platyrhynchus* – see photos opposite), one on the southern island of Novaya Zemlya (and possibly the New Siberian Islands – *R. t. pearsoni*), and the third on some northern Canadian islands (*R. t. pearyi*). In the case of the last two, interbreeding with mainland animals has occurred. This has meant that the pure-bred Novaya Zemlya subspecies is now considered lost, while the pure-bred Canadian island form (Peary Caribou – see Photo 243c) now probably exists only on the Queen Elizabeth Islands. The Svalbard Reindeer remains pure-bred as it is well-separated from mainland forms. All three island forms are smaller than mainland animals, with shorter legs. They are also more solitary, forming small herds or, often, being seen alone. These Reindeer are almost white in winter. The Reindeer of Kamchatka (*R. t. phylarchus*) are much larger than other subspecies, up to 50% bigger than nominate animals.

Reindeer were domesticated in Eurasia (this itself giving rise to a number of perceived subspecies) and truly wild animals are now found only in Norway, Russia (including the southern island of Novaya Zemlya and (introduced) Wrangel Island), though there are feral herds in other parts of Scandinavia. There are also feral Reindeer on Iceland. Caribou were never domesticated in North America and huge herds still cross the tundra on annual migrations. Caribou breed throughout Alaska, mainland Canada and the Arctic islands.

→Caribou (and wild Eurasian Reindeer) use traditional calving grounds, some herds of Caribou travelling more than 5,000km annually between these and winter feeding grounds. The migrations are in part a search for new browse, but also an escape from the predators that feed on the big herds, Wolves and Brown Bears rarely following the Caribou onto the tundra. (There are tundra bears and tundra wolves, but there are far fewer of them.)

On migration north the females forge ahead of the males, the later feeding on the richer southern tundra to put on weight for the rut. The females' progress is so urgent – the animals may travel more than 100km each day (though 25–65km is more usual, maintaining a steady pace of about 7km/h: the animals can run at up to 80km/h) – that if they give birth early they may abandon a calf that cannot keep up. For calves (normally one, rarely twins) the first hour is critical as it must dry (if the weather) is good, or be licked dry by the female to prevent it freezing to death. Once dried, the calf can walk within about an hour of birth, and can run several kilometres by the time it is two hours old, sustained by its mother's milk which is 20% fat and among the richest of any terrestrial mammal. Reindeer are good swimmers and will cross rivers: the Bathurst Caribou herd crosses Bathurst Inlet twice each year, a swim of 10km.

Famously, the animals feed on Reindeer Moss, actually a lichen rather than a moss, principally *Cladonia rangiferina*, but other species as well, these being located under snow by smell. But the lichens are mainly winter forage, spring's harvest allowing a greater range of foods, including fungi, green vegetation (plants and shrubs), seaweed and even, though rarely, lemming in 'lemming years'. They may also consume their own shed antlers to obtain the calcium for the new growth.

Observers of Caribou in spring/summer will note that the animals often rest on snowfields or seek out windy spots in an effort to avoid mosquitoes which otherwise make life almost unbearable, some animals being seen to run and jump frantically, stamping and head shaking in an effort to lose the clouds of insects. The Caribou are also preyed on by Warble Flies and Nose Bot Flies, the sight of either arousing deep fear in the animal as they are both parasitic.

a Nenet people of Russia's Yamal Peninsula moving their domesticated Reindeer herd to a wintering area.
b, c Wild Reindeer, Norway.
*d, e Svalbard Reindeer (*R. t. platyrhynchus*).*

Elk/Moose

The different names of this animal causes much more confusion than the Reindeer/Caribou pairing. The North American name, Moose, derives from an Algonquin name meaning 'twigeater', while the European name is from the Germanic *elch*. The confusion arises because in North America the name 'Elk' is given to another animal – *Cervus canadensis*, also known as the Wapiti, and very similar to the Eurasian Red Deer *Cervus elaphus*. In what follows 'Moose' will be used throughout, except in photo captions.

Moose are herbivorous, taking shoots and twigs of shrubs and trees (chiefly pine of the latter), green vegetation and aquatic plants. While the latter cover a range of species, the animals particularly seek Potamogetan pondweeds as these are rich in sodium which is in short supply in winter forage. Moose often submerge their heads to browse on these, and have also been known to dive to 4m to reach them. Moose go into water not only to feed, but to cool their stomachs. As with other ruminants, Moose have a multi-chambered stomach to assist in dealing with plant cellulose. Moose have four chambers, and the large fermentation pit this represents in such a large animal is a significant heat source. Unable to sweat, Moose stand in water to cool off. Moose seek out roads, both because travelling snow-ploughed roads made travelling easier than the deep snow of forests, but lick the salt that historically was used to aid motoring. Moose have long legs and in many vehicle collisions the animal's body is above bonnet-level. At any reasonable speed the result is a 500+kg animal crashing through the windscreen: Moose collisions are a significant cause of death on Alaskan roads, and the use of salt on Scandinavian roads has been banned to prevent such incidents (and collisions with Reindeer which also lick salt). Moose are also dangerous in other situations: dog owners and walkers encountering females with calves need to beware of an irate Moose rearing up on its hindlegs and lashing out with front hooves: the hooves are sharp and arriving with the force of full-grown Moose have been known to kill. In Scandinavia Moose consume apples in autumn. Enjoyed in quantity when over-ripe and fermenting, this causes the occasional Moose to become belligerently drunk, local papers running stories of animals menacing walkers and cyclists. As Moose are fast, this danger – indeed any close encounter – should not be underestimated: Moose can reach 55km/h. They can swim at 10km/h: there are stories of swimming Moose overtaking a canoe in which two fearful men were paddling frantically. This tale may not be apocryphal.

Moose (Elk) *Alces alces*

The world's largest extant deer, and apart from extreme specimens of the bears, the largest terrestrial Arctic mammal. (Bison (*Bison bison*) are larger in the southern part of the Moose's Nearctic range.) Adults are dark brown or dark grey-brown, the underparts paler, the legs pale grey-white or grey-brown. The tail is short, and as with the Reindeer, of little value in combating mosquitoes. The snout is bulbous and a large pendant 'bell' hangs from the throat. The bell's function is unknown, but bulls spatter it with urine-soaked mud during the rut. Male Elk have enormous antlers, which are usually palmate with a number of tines, though regular 'deer-like' antlers have been seen. In some large males the antlers can weigh 30kg and have a spread of 2m. Moose have poor eyesight, but exceptional hearing and studies have shown that the palmate antlers act as parabolic receivers, the Moose's ears moving to accept reflections from the antlers, enhancing the sound. As with Reindeer, during the rut male Moose bellow to establish hierarchy, but will also clash heads in trials of strength: there are tales of antlers becoming enmeshed, both animals dying of starvation.

Moose breed throughout the Arctic, though they are absent from Greenland, Iceland and all the Arctic islands. In Russia Elk do not breed east of the Ob River. Although essentially a boreal species, Elk are also found on tundra if foraging opportunities are good. Resident, but may move south by up to 150km in search of better winter forage.

a Female Moose and calves.
b Young bull Moose.
c Female Moose feeding on aquatic plants.
d This female Moose had been fitted with a tag in the Denali National park to observe behaviour and movements.
e, f Bull Eurasian Elks.

Musk Ox

With their huge heads and shoulder muscles, and long straggling hair that extends almost to the ground, hiding most of the short legs, Musk Oxen cannot be confused with any other species. They look primitive and comical, but are actually neither, the latter most definitely not, as their natural wariness means that if approached too closely they will inevitably charge the observer. Though best left at a distance, it is worth remembering that the heavy head and short legs mean the animals are unbalanced when running downhill, so approaches should be uphill with an expectation that a sharp turn and fast run downhill will be necessary. Downhill charges are short, terminated when the observer is seen to be fleeing. Fleeing uphill is not recommended: being caught by an irate Musk Ox will probably result in very serious injury: the animals (males may weigh up to 600kg) are known to toss wolves into the air if attacked and then to stamp on the grounded wolf.

The name derives from the habit of spraying urine onto their long guard hairs, this resulting in a musky smell. The smell may also serve a purpose in the rut, though males will also bellow to define hierarchy. Close matched bulls will take up aggressive stances: if one will not back down then they may charge each other, clashing heads at up to 40km/h. The animals live in social groups of 10-50 animals (size depending on available forage) with a dominant male, his harem of females, with calves and immature animals.

The massive head is also used to break through ice to reach winter forage and is also useful in defence against wolves. When attacked, Musk Oxen form a defensive circle with the calves at the centre. Each animal faces outwards, this presenting the pack with a ring of massive horns and head bosses, a blow from which would inflict significant, perhaps fatal, injury to an individual Wolf. As long as the oxen maintain the circle they are almost invulnerable. Stand-offs lasting several hours have been observed, though usually the Wolves give up and leave if attempts to break the circle fail. The Wolves make occasional forays forward, seeking to make the herd break up and run. Even if the Musk Oxen do make a run for it, they can foil the Wolves as long as they remain a cohesive unit. But sometimes the Wolves run into the herd, biting at flanks and seeking out a young or weak animal. Once an individual has been isolated the pack concentrates on it: the outcome is then a foregone conclusion. The circle is a good adaptation to Wolf predation, but is useless against a man with a rifle, and in several places where the animals were present, hunting led to local extinctions. Such slaughters in the Canadian Arctic led to the government protecting the species in 1917, and was the primary reason for the creation of the Thelon Game Sanctuary in 1927.

Musk Ox *Ovibos moschatus*

Although they belong to the same family (Bovidae) as cattle, Musk Oxen are not true oxen: they are more closely related to goats and sheep. Adult unmistakable, with its long straggling guard hair (individual hairs can be 70cm long), often long enough to cover the short legs. The pelage is dark brown or dark grey-brown, but with a paler saddle. The muzzle and legs are also paler. The fine underfur is claimed by many to provide the finest wool in the world. Known by the Inuit word *qiviut*, it is highly prized for the manufacture of (very expensive) small items of clothing, chiefly scarves. The upturned horns arise from each end of a central boss (up to 10cm thick) across the top of the head. The structure is formidable and is a significant fraction of the animal's total weight. The hooves are broad, with a sharp outer rim and a softer inner one, giving the Musk Ox surprising agility on rock. In females, the head boss is not continuous, a fur patch separating the horns. Several subspecies have been proposed, but re-introductions have made separation of these difficult.

Musk Oxen breed in western and northern Alaska (though these are introduced animals originally released near Fairbanks, then on Nunivak Island in 1935 as the native stock was eliminated by hunting - to provide meat for whalers - in the 1860s), on the northern rim of mainland Canada west of Hudson Bay, and on western and northern Canadian Arctic islands. Absent from Baffin Island, but a small herd breeds near Ungava Bay in Quebec. Also breeds in north-east and north Greenland, and has been re-introduced to north-west and western Greenland. Absent from Iceland. Was introduced to Svalbard, where it almost certainly never occurred naturally, but this population is now extinct. Has been introduced to Scandinavia and (reintroduced) to Russia, but populations are small.

Sheep

There are no true Arctic sheep, but Dall's and Snow sheep breed at the Arctic fringe. Historically, both were considered subspecies of the North American Bighorn Sheep (*Ovis canadensis*) but each are now considered to have full species status. As with Musk Ox, the sheep have long outer guard hairs (kemps) which are hollow, this being an efficient way of conserving heat. Beneath the guard hairs is an insulating fleece. Both sexes grow horns, but these differ significantly, those of ewes being spike-like, while those of males are long and curved. The ribbed rings which form the horns allow the animals to be aged. The sexes live sepasfrately, ewes and lambs staying in areas where forage is poor, which tends to limit predator numbers, while rams live in areas of better forage, trading the higher risk of predation with the weight gain which affords status and the right to mate.

Rams move to ewe areas for mating. In Dall's Sheep the rams do not hold territories or harems. Instead they move among the ewes until discovering an oestrus ewe. The ram then defends the ewe against rivals until mating occurs. The ram then moves in search of another ewe. Defence may involve butting heads, each ram rearing up on its hindlegs prior to the butt to gain momentum. On wind-free days the crash of heads can often be heard over a distance of several kilometres. Very little is known of the mating behaviour of Snow Sheep (indeed, very little is known about any aspect of the poorly studied species), but it is assumed to be similar to Dall's Sheep.

Dall's Sheep (White Sheep, Thin-horn Sheep) *Ovis dalli*

Adults are white (though Stone's Sheep, *O. d. stonei*, a subspecies from southern Yukon, is grey-brown). The hooves are black, the tail very short. The horns of rams curve back from the forehead, and are then strongly downcurved with forward-pointing tips: the horns are up to 100cm long. The horns of ewes are smaller and straighter, and up to 25cm in length.

Dall's Sheep breeds in upland areas of Alaska and Yukon, but rarely to the north coast. In general they prefer steep ground.

Snow Sheep *Ovis nivicola*

Comparable in size to Dall's Sheep, rams being up to 100kg. Adults vary in colour. They are usually brown with paler underparts and rump patch. The rump patch is often white and may spread onto the lower back. The forelegs are usually darker. The horns are similar in size and form to those of Dall's Sheep.

Snow Sheep breed in upland areas of eastern Siberia, including eastern Yakutia, Chukotka and Kamchatka. Three subspecies are recognised by some authorities – Yakutian Snow Sheep (*O. n. alleni*), the Koriak Snow Sheep (*O. n. koriakorum*) of northern Kamchatka and southern Chukotka, and the nominate of southern Kamchatka. However, the suggested differences in colour seem to be encompassed by the differences seen between individuals.

a Dall's Sheep ewe and lamb, southern Alaska.
b Dall's Sheep ram, Denali National Park.
c, d Snow Sheep rams, southern Kamchatka.

Carnivores

The large carnivores, Wolf, Brown Bear and the area's most iconic animal, the Polar Bear, are at the top of the 'must-see' list for most Arctic travellers. But carnivores are a large and diverse group of mammals, including the pinnipeds (aquatic, and almost exclusively marine) as well as the fissipeds, the terrestrial group. Arctic representatives of the latter group includes the largest and smallest of Earth's carnivores, the Polar Bear and the (Least) Weasel – one weighing up to 800kg, the other just 25g.

So different are the many forms of carnivores that only a few general comments on physiology can be made. There are two obvious skeletal differences if compared to human anatomy. There is no large clavicle, and a penis bone is present in almost all species. The function of the clavicle in humans and other primates is to allow sideways movement of the arms. For carnivores, whose main hunting strategy is to run down prey, such a movement would be disadvantageous, and they have a smaller clavicle, free at each end, which allows only a forward and backward movement, ideal for fast pursuit. The *baculum* or *os penis* bone of carnivores prolongs copulation and is assumed to have evolved because in general copulation stimulates ovulation. The 'copulatory tie' seen in mating canids is not thought to play the same role, though it clearly has that effect. The tie, which can last for up to 30 minutes, is almost certainly related to sperm competition: longer coupling helps ensure that a rival male's sperm will not subsequently fertilise the female. Carnivores also have a unique and diagnostic dentition. In all carnivores (even in the only species that has become secondarily herbivorous, the Giant Panda *Ailuropoda melanoleuca*), the last upper premolar and first lower molar on each side of the jaw have high cusps and sharp tips. These teeth, carnassials, allow flesh to be sheared.

In addition to the physical attributes for hunting terrestrial carnivores also have well-developed senses, sight, hearing and smell all being important. These are also used in communication, the animals marking their presence by urinating and defecating at strategic points throughout their territories.

Canids

Canids (dogs) are noted for pack formation, the pack normally comprising a monogamous 'alpha' pair which breeds, and a number of family members that help with hunting and pup-rearing. However, while this is the normal society for Wolves and Coyotes, Red and Arctic Foxes are more solitary animals. Family groups have been noted in these foxes, but solitary animals or pairs are more likely to be seen in both species. The number of pups produced is dependent on prey density: Arctic Fox numbers increase rapidly in 'lemming years', but the animals may not breed if rodent numbers are low. Many foxes will starve during their first year if the lemming population crashes. In general canids kill their prey by grabbing the neck or nose, and shaking their heads violently to break or dislocate the neck or spinal cord.

Wolf *Canis lupus*

Status: CITES: App. II (but wolves of the Indian sub-continent are App. I).

The largest wild dog. The first animal domesticated by humans, perhaps as many as 50,000 years ago, and certainly by 14,000 years ago. Despite this ancient association, the human view of wolves has undergone a significant change over time. Once the familiar of shamans wolves were revered by early northern dwellers: the Vikings also revered the animal for its speed, stamina and skills as a hunter, Viking chieftains often taking the name Ulf to indicate their prowess in battle. But attitudes changed, the animals becoming feared and loathed, a view that is still prevalent in most communities which share space with wolves.

a Wolf pack, northern Fennoscandia.
b, c Alpha pair (female on left), Ellesmere Island. As noted on p238, Arctic Hares do not change colour in the short northern summer, and neither do their main predators.
d It is spring in Canada's Barren Lands, and this Tundra Wolf is starting to moult to summer pelage.
e Adult Wolf, northern Fennoscandia.

But by the early medieval period things had changed, the Wolf becoming an evil presence, the change reflected in a host of folkloric tales in which the animal is represented as deceitful, greedy, ferocious and, most significantly of all, dangerous. In many European countries a bounty was placed on Wolves: the last Wolf was killed in Sweden in 1966, and in Norway in 1973. They were almost exterminated in Finland too, but recolonisation from Russia prevented total extinction. As attitudes changed, the recolonisation of Sweden and Norway from Finland was officially welcomed, though the animals still face illegal killing by farmers convinced that the Wolf represents a serious threat to livestock, and by people who, fearing attacks on children and the vulnerable, feel that such large carnivores should not be allowed to co-exist in a civilised society.

Unfortunately, Europeans settlers to North America took their prejudices with them, and Wolves were persecuted mercilessly. Only relatively recently has the change in public attitude allowed the Wolf to be re-introduced to the Yellowstone National Park. However, in many places outside National Parks the Wolf is still under threat. Interestingly, the Inuit have a curious relationship with both dogs and wolves. Sedna, one of the main spirits of the Inuit world, the mother of all sea animals, was reputed to have mated with a dog, some of her children being dog-like, creatures that were the ancestors of Indians and white men. The Iniut believed that dogs had no souls as they lived in too close an association with people. They noted the hunting technique of the wolf, seeking out sick or injured animals so that the kill was less dangerous for the hunter. It seems they both admired the wolf, but were contemptuous of it, perhaps because usually the Inuit were not fortunate enough to be able to concentrate on easier prey. When policemen were first introduced into Inuit societies by the Canadian government they rapidly acquired the name *amakro* (wolf) because they lurked in the background observing, watching for mistakes.

Adult wolves are usually grey, but this can vary (particularly in North American animals) from almost black to white. White Wolves are found in northern Canada and the Canadian Arctic islands, in Greenland and areas of Arctic Russia. Wolves are social animals living in packs, which are cohesive, extended family groups. Packs form when a mature male and mature female from two packs meet, form a breeding pair and establish a viable territory. Pack sizes vary with prey size, prey density and Wolf mortality (rabies and other fatal diseases being endemic in Wolf populations). A pack can contain as many as 30 animals (with numbers as high as 60 recorded). Pack sizes in general are much smaller for High Arctic Wolves: on occasions a 'pack' there might consist only of a male and female. The pack is dominated by the 'alpha' pair. In general the alpha animals are the only pack members that breed, the non-breeding members being earlier offspring who help with hunting and cub-rearing. The female breeds annually, but in the High Arctic breeding may not occur if resources are inadequate. In those cases the foetuses are resorbed.

The howl of the Wolf is one of the most evocative sounds of the wilderness. Howling helps the pack to reform after a dispersive hunt. In the Arctic the prey is often Reindeer, though animals as large as Moose may be taken. In the High Arctic, Wolves take rodents and hares and, if the pack is large enough or the prey is weak, they may take Musk Ox, usually an old or sick animal, or perhaps a male which has left his herd after being replaced by a younger male.

Wolves breed throughout the Arctic, but are absent from Iceland and Svalbard. Breeding on Russia's Arctic islands is conjectural as the species has not been well-studied in those areas.

Coyote *Canis latrans*

Adults vary from red-brown to grey or even dark grey or black. Coyotes are essentially subarctic, but they are opportunistic feeders and so may venture north if resources allow. They breed across the United States and southern Canada. They have been seen on Alaska's North Slope and around the Mackenzie delta, though the timberline is the more normal limit.

a, c Wolves, northern Scandinavia.
b Opportunist meets opportunist, Raven and Wolf, northern Scandinavia.
d Coyote.
e Tundra Wolf, Canadian Barren Lands.

c

e

Arctic Fox *Alopex lagopus*

Apart from the Polar Bear, the Arctic Fox is the most specialised terrestrial Arctic mammal. The ears, muzzle and legs are small to reduce heat loss, and the tail is large and bushy, allowing it to be curled arond the body as a blanket. Though usually seen on land hunting rodents, the foxes are often seen on sea ice in the winter, where they account for about 25% of the annual kill of seal pups, and follow Polar Bears to feed on the remnants of bear kills (and on bear faeces in extremis). The foxes also follow Wolves on land for the same purpose, and regularly patrol bird nesting cliffs to collect eggs and chicks from poorly sited nests, or to take young birds whose first flights (or glides in the case of some auks) end in disaster. In 'lemming years' large numbers of cubs wil be born and raised, but winter mortality and lack of available prey the following year will rapidly reduce numbers.

Adults have two colour forms. In one the summer pelage is grey-brown above, paler below and, usually, on the tail. In the winter this form turns white. The second form, the so-called blue fox, is dark chocolate brown or dark blue-grey in summer, with a paler tail, and pale blue-grey in winter. In general white-morph foxes are continental, blue foxes being found on islands, but interbreeding of the forms occurs and litters may comprise both colour forms. In such litters cubs are born dark brown, the two colour morphs not being identifiable until they are about two weeks of age.

The luxurious winter coat of the Arctic Fox – so thick it makes the animal look much bigger than it actually is – is such a good insulator that a body temperature of 40°C can be maintained without the need for shivering down to ambient temperatures of about -60°C. Unfortunately, the coat was so prized that the animals were relentlessly hunted both in Russia and Canada (see the Introduction for further details). As noted there, the relentless drive for furs meant other animals were killed, 'collateral damage' in traps set for foxes, and the introduction of foxes onto many Aleutian islands was disastrous for ground nesting birds. But despite the staggering numbers of foxes that were killed the species survived, a tribute to its fecundity and its ability to thrive in an extreme environments. Changes in fashion and, particularly, in the attitude towards the wearing of real fur in western countries reduced the take of hunting. Hunting still continues, but at a lower level and the IUCN status of the animal is 'Least Concern'.

Arctic Foxes do not form packs, being seen a solitary animals or as pairs in the breeding season. On occasions small groups of animals may be seen if they congregate at a food source, for instance a beached whale. Fox litters are born in a den excavated in the tundra.

Arctic Foxes breed throughout the Arctic, including islands as far north as Ellesmere and Svalbard. However, the species was been hunted to extinction on Jan Mayen. It is possible that the island might be recolonised if sea ice allows animals to reach it again. Icelandic animals have been persecuted since human colonisation of the island, but particularly after the growth of Common Eider 'farming' for eiderdown when a bounty was paid for fox tails. The retreat of sea ice has meant that new foxes have not arrived to breed with what remained of the indigenous population, though fur-farm escapees has aided gene diversity. Though still uncommon, the foxes are no longer the very rare species they had become.

a Two Arctic Foxes feeding on a Musk Ox carcass in the Canadian Barren Lands. It is winter, a time when the foxes tend to be solitary, but the bountiful supply of food at the carcass has overcome their desire to avoid conspecifics.

b Blue Fox on the Pribilofs. The fox is searching the beach where Northern Fur Seals have given birth, on the lookout for a stillborn pup, a small pup with a distracted mother or afterbirth.

c Arctic Fox at the remains of the Polar Bear kill, Svalbard sea ice.

d Blue Fox cub.

e Normal morph winter pelage.

f, g This litter in north-east Greenland includes both colour morphs. The cubs are about a month old.

Red Fox *Vulpus vulpus*

The Red Fox is the most widespread and abundant of all carnivores, breeding across Eurasia, North America and north Africa. Most adults are red-brown with a bushy tail. The tail is often white-tipped, the ears usually black-tipped. The feet and legs are often darker than the body colour. Though most foxes conform to this colour pattern, there are exceptions, particularly in North America, where there are paler animals, and where certain other colour forms common enough to have been given specific names. The 'Cross Fox' (relatively abundant in Canada) is darker with much darker shoulders and back, these darker sections of fur forming the cross of the name. The 'Silver Fox' is very dark, often black, with white-tipped guard hairs which give the animal a silver sheen in much the same way that gives 'Grizzly' Bears their name. There is also a blue-grey form similar to the blue Arctic Fox: in the United States this morph is often given the inelegant name 'Bastard Fox'.

Red Fox breeds across the Arctic, but are confined to southerly latitudes, being absent from Greenland and high-latitude Arctic islands (i.e. Svalbard, Russian islands apart from the southern island of Novaya Zemlya, and Canada's northern islands). Also absent from Iceland.

In areas where both Red and Arctic Foxes occur, the larger Red will drive the Arctic out, outcompeting it and/or predating the smaller animal. As the Red cannot thrive in areas of sparse prey, far-northern latitudes favour the Arctic Fox. The northern limit of the Arctic Fox is therefore defined by the availability of food (and of den sites), while the southern limit is defined by the northern limit of the Red Fox. As the Earth's climate warms and species move north it is likely that eventually Red Foxes will oust Arctic Foxes from mainland Eurasia and North America, the smaller fox being confined to high latitude islands that the larger fox cannot reach.

Cats

The Felidae or cats are the most carnivorous of the Carnivora (with the probable exception of the Polar Bear). They are superbly adapted for the role, having exceptional eyesight, hearing and smell, phenomenal agility and (in all bar one) sheathed claws, the sheath maintaining the claws in good condition by avoiding wear. Although primarily warm-climate predators, some cats do live in cold conditions. The Siberian Tiger *Panthera tigris altaica* inhabits southern Siberia, as does the critically endangered Amur Leopard *Panthera pardus amurensis*, while the Snow Leopard *Panthera uncia* lives high among the snows of the Himalayas. However, no cats are truly Arctic. The Old and New World lynxes are primarily boreal hunters, but they may be seen beyond the timberline. Lynxes take rodents, hares, young deer and birds.

Eurasian Lynx *Lynx lynx*
Status: CITES: Appendix II.
Adults are stocky with a yellow-brown upper pelage, pale or cream below. There are usually dark spots and blotches on the body and the fronts of the long legs. The face is streaked black. The tail is short and black-tipped.

Eurasian Lynx breed in central Scandinavia (where they are rare) and across Russia to Kamchatka, though only to the north coast near the White Sea and the Lena delta. In Kamchatka the animals take Snow Sheep.

Canadian Lynx *Lynx canadensis*
Status: CITES: Appendix II.
Adults are as the Eurasian Lynx but are overall grey or grey-brown, with denser dark flecking on the upper parts.

Canadian Lynx breed in Alaska (but not on the Yukon–Kuskokwin delta or the southern Alaska Peninsula, and are rare in the north), the Yukon, North-West Territories, Quebec and Labrador, but not to the northern coast.

> *a-d Red Foxes, northern Scandinavia.*
> *e Red Fox, northern Canada.*
> *f Canadian Lynx.*
> *g Eurasian Lynx.*

Bears

The three northern species of bears – the Polar, Brown and Black Bears – are at least double the weight of the largest of their southern cousins, size and a low metabolic rate being adaptations for surviving the northern winter. The northern bears are large and powerful creatures, the Brown and Polar Bears being the largest terrestrial carnivores on Earth.

Both Brown and Black Bears exhibit periods of winter torpor that require the laying down of fat reserves. Pregnant female Polar Bear undergo winter torpor in the maternity den. Male Polar Bears do not have such periods: they will dig a den in which to rest during periods of bad weather. Yet despite not having a winter torpor period, male Polar Bears have retained the size of the Brown Bear from which it has relatively recently evolved. Exactly when the split occurred is still debated. It is assumed that a population of Brown Bears were isolated from conspecifics during an ice age and evolved to cater for the icy habitat they found themselves in. The split would therefore have occurred during the last million years, and very probably within the last 250,000 years, but perhaps as recently as 100,000 years ago. In support of a 'recent' split, the two bears can mate successfully, hybrids being known in both captive and wild bears, though the latter are rare. Although hybrids had been suspected, following the shooting of a hybrid near Sachs Harbour on Banks Island in 2006 DNA testing showed that the bear was indeed a hybrid, a male Brown Bear having mated with a female Polar Bear. The finding meant the hunter escaped a fine/jail time for the illegal killing of a Brown Bear. In 2010 another hybrid was confirmed. Shot on Victoria Island, DNA confirmed that this animal was fathered by a Brown Bear, but that its mother was herself a Brown-Polar hybrid. The offspring of Brown-Polar matings are therefore fertile.

The northern bears share a very similar dentition. Brown and Black Bears are omnivorous, and they have molars with broad, rounded cusps suitable for grinding plant material, and a diastema – a gap between the canines and the molars created by absent (or vestigial) premolars – which is used for stripping bark. Each of these adaptations for omnivory is a disadvantage for the wholly carnivorous Polar Bear.

Bears do not form long-lived bonds, mating occurring when a male bear encounters a female in mating condition (which helps explain hybrids, Brown Bears moving north as the Arctic warms). Male Polar Bears, which do not hold territories, attempt to keep a female away from other males while she is receptive. Male Brown and Black Bears do hold territories, and these usually overlap the territories of several females, all of which the male will mate with. However, females will also mate with other males they encounter while receptive. Ovulation in bears is stimulated by copulation, the fertilized egg developing to a blastocyst whose implantation is delayed. It is believed that delayed implantation evolved in Brown and Black Bears to allow females to focus on laying down fat reserves in late summer and autumn, rather than cub-rearing. Courtship and mating take place in late spring, the blastocyst being implanted during autumn. Birth occurs in the maternity den during the female's winter torpor. At birth the altricial cubs are very small. Mother bears lose up to 40% of their body weight during torpor (and up to 50% in some female Polar Bears). As a consequence, the female's first requirement when leaving the maternity den with her newborn cubs is to go in search of prey (Photo 261a, the maternity den was on Edgeøya, a small island to the east of Spitsbergen, the mother and cubs walking past an iceberg calved in northern Svalbard and frozen in when the sea ice formed).

Polar Bear *Ursus maritimus*

Status: IUCN: Vulnerable (the population is poorly understood, as are Polar Bear population dynamics. Climate change has increased the number of pathogens in the Arctic and the number of persistent organic pollutants is also increasing. Each will likely have a detrimental effect on the bears. As ice coverage reduces, gas and oil exploration and production may also have an adverse effect). CITES: Appendix II.

The symbol of the Arctic, not only for the traveller but also for the Inuit, who fear and admire the bear in equal measures. Polar Bears are cream or pale yellow (Photo 261b), rather than white (though when seen against a dark background or in flat light (Photo 261c) they certainly do look white). The pelage comprises a thick underfur up to 5cm long with guard hairs that grow to 15cm. A layer of blubber 5–15cm thick lies beneath the skin. So good is the bear's insulation that they are more in danger of overheating than of hypothermia, particularly

if they are running. In short bursts they can achieve 40km/h, but such activity rapidly causes overheating. Hot bears lie on their backs and expose the soles of the feet to cool down, an endearing posture. In warm weather the bears will excavate dens in the snow to keep cool. In winter male bears also excavate dens to escape prolonged bad weather. In water the bears travel at about 2–3km/h and can cover great distances (Photo 263f), though tales of bears up to 100km from shore almost certainly refer to individuals carried on an ice flow that subsequently melted.

An adult Polar Bear has a characteristic 'roman' nose and lacks the shoulder muscle hump of a Brown Bear. The nose and lips are black, the ears are small, but the animal has large feet to aid walking on snow (Photo 263b). The forepaws are also used as paddles by the swimming bear; the back legs are not used in water and trail behind the animal.

Polar Bears primarily hunt seals, chiefly the Ringed Seal. Female seals give birth to their pups in dens above the sea ice, the pups protected by a layer of compacted snow into which the female digs. When a bear detects a den it rises on its hind legs and crashes down through the den roof. If the bear is lucky it seizes the pup or blocks the escape hole into the ocean. If the bear is unlucky it fails to break through or misses the pup, which escapes (Photo 263a, the female's newborn cubs watching attentively as their mother attempts to secure a Ringed Seal). Bears that fail to break through immediately will often try again, but the chances of catching a pup are much reduced if more than one attempt is necessary. On average, a bear succeeds in catching a pup in about one in three attempts; if multiple attempts are needed to break through the roof this percentage falls rapidly. The technique is also used for hunting resting adult seals.

Bears also wait at breathing holes catching seals as they surface. They will also attack Belugas stranded in leads in the sea ice, clawing at a whale each time it surfaces until the animal becomes exhausted by blood loss and anxiety and can be hauled on to the ice. Polar Bears will also search among hauled-out walruses for pups. In the sea a walrus is a match for a bear, and may even inflict fatal damage with its tusks on land, but large walruses are not very mobile and the nimble bear can occasionally steal a new-born pup from an irate, but helpless, mother (Photo 263d shows a bear hopefully searching among walrus on Wrangel Island; Photo 263e shows the remains of a walrus pup caught by a bear on a small island near the entrance to Fury and Hecla Strait at the northern end of Canada's Foxe Basin). There are many stories of the cunning of bears; these tell of bears that excavate the breathing holes of seals to make capture easier, then shield the hole with their heads so the seal does not observe a change in light level, and of bears hunting with a paw over their noses (the only black mark on them). Though apparently easy to dismiss, there is historical evidence to support claims of such crafty prey-capture techniques. Clements Markham, a midshipman on the British ship Assistance, one of the four ships of Horatio Austin's Franklin search expedition of 1850–51, reported in a letter to his father that he watched a Polar Bear 'swimming across a lane of water [and] pushing a large piece of ice before him. Landing on a floe he advanced stealthily toward a couple of seals basking in the sun at some little distance, still holding the ice in front of him to hide his black muzzle.' And there is a credible story of a bear stalking a seal in water by swimming towards it when the seal surfaced, then floating, motionless when the seal submerged. Seal eyes are at their best underwater so this strategy allowed the seal to believe the bear was a piece of floating ice. The strategy was apparently successful.

Polar Bears are found throughout the Arctic and have been seen as far south as Iceland, mainland Scandinavia and even Japan's northern island, Hokkaido, though with sea ice coverage now in full retreat as a consequence of global warming it is unlikely these destinations will see bears again. Female bears regularly den in James Bay (at the same latitude as London). These bears, forced to live on land when the sea ice retreats, may eat berries and plants while waiting for the sea ice to reform, though many actually fast for many months. The sea ice of western Hudson Bay forms first at Cape Churchill, explaining why the bears (and bear watchers) congregate there. When the sea ice forms, the bears disappear.

Brown Bear *Ursus arctos*

Status: CITES: Appendix II (though the subspecies in Bhutan, China, Mexico and Mongolia are App. I).

Arguably the largest bear. Weights in excess of 1,000kg have been claimed for Kodiak Brown Bears (which exceeds the largest-known Polar Bear weight). Similar weights may also be achieved by the bears of Kamchatka, which, like those of Kodiak, feed on spawning salmon and therefore put on huge fat reserves before winter torpor. Because of geothermal energy some rivers and lakes in Kamchatka remain unfrozen during the winter and salmon may make late spawning runs, the bears delaying or avoiding torpor to feed on them. However, in general both Brown and Black Bears enter a state of torpor, a strategy forced on them by a lack of winter food. In each case the heart rate reduces (from 40–70 beats/minute to c.10 beats/minute in the case of the Brown Bear), though body temperature is maintained just a few degrees below normal. This means that the bears can swiftly rouse themselves if danger threatens, this being particularly important for females with cubs. The bears are able to survive for up to six months not only without eating or drinking, but also without urinating or defecating, an ability not shared by any other mammal.

Brown Bears are uniformly dark brown, though some individuals are paler (and may even be very pale or grey-brown), while others may be almost black. Some bears have paler tips to the guard hairs, giving them a 'grizzled' appearance (hence the name Grizzly Bear). However, some Grizzlies may be so pale overall that this is barely discernible. The muscle-hump at the shoulders is prominent in many individuals. In general Brown Bears have a concave head profile i.e. a 'dish' face. Brown Bears are omnivorous, feeding chiefly on plant material but taking mammals opportunistically (they will sometimes excavate rodent burrows) and will occasionally chase down sick animals as large as Moose, or young animals, and some feed extensively on spawning salmon.

Brown Bears breed throughout the Arctic (the majority of the population is in Russia), but they are absent from Greenland and in general are absent from all Arctic islands. A possible exception is Banks Island where there are credible stories of Brown Bear sightings and where it is now known (see the introduction to the section on Bears) that mating between a male Brown Bear and female Polar Bear occurred. Since a later hybrid was found on Victoria Island it may be that a warming Arctic will see a more general northward advance of the species.

Black Bear *Ursus americanus*

Status: CITES: Appendix II.

Adult bears have a range of pelage colours. Black is, not surprisingly, the usual colour, but the Kermode Bear of west central British Columbia (chiefly found on offshore islands), is white, the Glacier Bear of Glacier Bay, Prince William Sound is blue-grey and the Cinnamon Bear of Alaska varies from pale cinnamon to dark brown.

Black Bears breed in the forests of Alaska and Canada, rarely being seen above the timberline. Black Bears are the most numerous of the bears, with some estimates suggesting that there are more Black Bears than the combined populations of all the other species put together. The population is apparently stable despite an annual mortality due to hunting of tens of thousands of animals. One study suggested that if a bear reaches the age of two, it stands a 90% chance of dying as a result of being shot, trapped or being struck by a vehicle. Human activity therefore represents the greatest risk of mortality to an individual bear, a chastening statistic.

a, b Brown Bears, Karelia, northern Finland.
c Juvenile Brown Bear, Chilkat River, south-east Alaska. Juveniles usually spend their first winter, after a winter birth in the maternity den, with their mothers learning how to fatten up for torpor and how to select a den. This yearling's mother came into season early, mated and evicted him. Having watched bears catch salmon on the river he eventually learned the technique. But it was now autumn and it is likely he did not survive the winter.
d Kamchatka Brown Bear, irate having been woken from a snooze in the sun.
e 'Grizzly' Brown Bear, Denali National Park.
f, g Black Bears. Although it is more satisfying to photograph the bears in natural surroundings, it is often easier to see them at the local dump.

Mustelids

The Mustelidae are perhaps the most diverse of the carnivore families, varying in size from the (Least) Weasel to the Sea Otter (the latter being more than 2,000 times heavier than the former). However, all share common features. The body is long and slender, the legs short, the shape ideal for pursuing prey in burrows (and seen even in mustelids that do not adopt this hunting technique – the Wolverine, for example, which is too large to do so). This shape, ideal for hunting small rodents, has the disadvantage of a large surface area-to-volume ratio, which is not optimal for reducing heat loss. As a consequence, the metabolic rate of the animals is high and they must hunt frequently, despite the high energy value of their diet. Some of the smaller female mustelids, when nursing young, must consume in excess of 60% of their body weight daily. One other feature common to the group is the shape of the head, which is flattened and wedge-shaped, another adaptation for hunting in burrows.

Size for size, the smaller mustelids are perhaps the most formidable of the carnivores, with Weasel and Stoat the champion hunters. It is possible that the absence of a significant Weasel population might actually be the initial cause of a 'lemming year', the lack of normal predation resulting in a rise in the number of first-litter young reaching sexual maturity, which then drives up the population sharply. In general mustelids kill their prey with a bite to the base of the skull, crushing the brain, or to the neck, severing the spinal cord. →

Wolverine *Gulo gulo*
Heavy animals with short legs, large paws and bushy tails. Adults are dark brown, with pale brown/golden stripes from behind the front leg across the lower flank to the rear, and across tail base. There are similar patches between the eyes and the ears (Photos 267a–c).

The Wolverine is the largest Palearctic mustelid (in the Nearctic the Sea Otter is, in general, larger). In Europe the animal is often called the glutton (this being the basis of the species' scientific name) because of its reputation for greed. It is known to take bait from traps set by fur trappers, and to eat trapped animals, these habits hardly endearing the species to people. It also has a reputation for being a cruel killer. The Sámi claim that if a satiated Wolverine catches a Reindeer it will gouge out its eyes so that the deer cannot move. The Wolverine will then, it is said, return when it is hungry to kill and feed on the still-fresh animal. In the absence of significant competition from Wolves this strategy might just work, but historically Wolves and Wolverines have shared ranges. More importantly, Wolverines invariably dismember large prey and cache sections in well-separated places. The Sámi tale is likely to be black propaganda against an occasional Reindeer predator.

Wolverines breed throughout the Arctic, but are absent from Greenland, Iceland, Svalbard, Russia's Arctic islands and some Canadian Arctic islands.

(Least) Weasel *Mustela nivalis*
Adults are chestnut brown above, white below, the delineation very distinct. The chestnut tail lacks the black tip of the Stoat (Photo 267d – this weasel was hunting Sand Martins in Finland). Winter pelage is entirely white.

Weasels breed throughout the Arctic, but are absent from Greenland, Iceland and the Arctic islands of Canada and Russia.

Stoat (Short-tailed Weasel) *Mustela erminea*
Coloured as the Least Weasel but with a distinct black tail tip (Photos 267e–g). The black tail tip is retained in the white winter pelage. Photo 267h is not a winter Stoat, but an albino – note the absence of a blacktail tip. Stoats are predated by raptors and some terrestrial mammals, and this animal, photographed on Canada's Victoria Island, has done well to survive to adulthood given its highly conspicuous colour. The North American name, Short-tailed Weasel, relates to the length of the tail relative to the Long-tailed Weasel (*Mustela frenata*) of the southern Canada and the USA.

Stoats breed wherever rodents are found. This includes Greenland, but not Iceland or Svalbard. Breeds on Canada's Arctic islands as far north as Ellesmere. Palearctic Stoats are, in general, boreal, and have not been observed to date on Russia's Arctic islands, but probably breed on islands with rodents. Nearctic Stoats are the most northerly mustelids.

(Eurasian) Otter *Lutra lutra*

Status: IUCN: Near Threatened (population decline as a consequence of habitat loss and pollution). CITES: Appendix I.

Adult is dark brown above, paler (even cream) below, the two colours not well delineated. On the face the pale colour extends to the lower ear. The tail is long, thick at the base and tapering to the tip, and uniformly dark brown. All the feet are fully webbed. The broad muzzle has many stiff hairs that are tactile and help locate prey in murky waters (Photos 269a-c).

Eurasian Otters breed in northern Scandinavia and across Russia to southern Chukotka and Kamchatka, but it is rarely found north of the Arctic boundary as defined in this book.

American River Otter (Northern River Otter) *Lontra canadensis*

Status: CITES: Appendix II.

Adults are almost identical to the Eurasian Otter, being rich dark brown above, pale brown or cream below, with a silvery sheen on the breast (Photos 269d and e).

More Arctic in its range than its Palearctic cousin, breeding throughout Alaska (though rare in the far north and absent from the Aleutian and Bering Sea islands), the Yukon, North-West Territories, around Hudson Bay and throughout Quebec and Labrador. However, the species is rarely found as far north as the coast.

American Mink *Neovison vison*

Adults have luxurious, lustrous dark brownish black fur, with a white patch on the lower lip. Feet and tail coloured as body fur. In Europe escapes from fur farms (which now breed over much of the continent) may be exotically coloured (e.g. cream and blue), but populations seem to be reverting to the natural colour (Photo 269f, Scandinavia, Photo 269g, North America).

American Mink breed throughout the Arctic, but are absent from Greenland and all Arctic islands (apart from Iceland, where a population has become established following fur farm escapes). Originally found in the Nearctic only, Palearctic animals are localised but spreading at a remarkable rate, and now occurs throughout Scandinavia and Russia. The presence of American Mink are considered to be influencing the survival of the Eurasian Mink (*Mustela lutreola*), an almost identical animal, but having a white upper lip and chin as well as the white lower lip. Interestingly, despite the morphological similarities, DNA analysis shows that the two species are not closely related. The species once bred across Scandinavia and western Europe, but is now extinct apart from small colonies in northern Spain and western France, and in eastern Europe including Russia, where it breeds around the southern shores of the White Sea and eastwards to the Ob River. The range is fragmented and the IUCN status of the species is 'Critically Endangered', with the population declining towards extinction as a result of historical overhunting, habitat loss and because the species is outcompeted by American Mink.

→Most mustelids are sexually dimorphic. This reaches an extreme in the Weasel, where the male is sometimes twice the weight of the female. This may be a response to prey availability, the two sizes meaning that males take larger prey and so do not compete with potential mates – but while this might be true of southern mustelids, the prey of most northern species is limited to the same range of rodents. An alternative theory suggests that females are smaller so that their energy needs are reduced, requiring them to catch fewer prey when they are lactating. Males could also be larger to increase the likelihood of their holding territories and competing for females.

In addition to the mustelid species covered here other, Arctic-fringe species may be seen by travellers. In the Nearctic the American Marten (*Martes americana*) breeds to the timberline, as does the Pine Marten (*Martes martes*) in the western Palearctic. In the eastern Palearctic the Sable has a patchy distribution across Siberia, but is relatively common in Kamchatka. In southern Alaska, the Aleutians, Commander Islands and south-eastern Kamchatka the endangered Sea Otter can still be seen in isolated pockets. Both the latter species are considered in more detail in the Introduction where the history of Arctic fur-trapping is explored.

Marine mammals

As the solubility of gases in water increases with decreasing temperature, Arctic waters are oxygen-rich. They are also nutrient-rich, due to the huge inflow of the rivers of Asian Russia and North America. The combination makes the Arctic seas highly productive in summer, attracting not only seabirds but marine mammals. While the pinnipeds that exploit this summer bounty remain in the Arctic year-long, moving with the ice edge, cetaceans travel to the area to feed in summer, then migrate to more productive waters during the winter.

The marine environment offers advantages to mammals in addition to a copious food supply. Freed from the major constraint imposed by gravity, marine mammals have fewer limits to size. Seals are restricted in size by the need to come ashore to give birth, but whales have overcome this restriction, and become massive as a result: Blue Whales (Photo 254a) are probably the largest animal ever to have lived on Earth. Giving birth in water would seem illogical for an air-breathing animal, but it has its advantages: female whales do not have to support the weight of their growing foetuses in the way that terrestrial mammals do, and so can carry larger young (within the obvious limitations of the birthing process) and the well-developed offspring are better able to seek air immediately, to overcome the problems of suckling underwater, and to survive the often hostile environment into which they have been born.

The buoyancy of water means that whale skeletons can be much less bulky. Honeycombed whale bones are remarkably light in comparison to those of heavy terrestrial mammals. The bones have a hard outer shell covering a sponge-like inner layer with numerous blood vessels and a marrow rich in oil: when whales were hunted, about 30% of the oil obtained from a carcass came from the bones. Though strong enough to act as anchors for the whale's huge muscles, some whale bones are so light that they float in water. Photo 255c shows Bowhead bones at the ritual site on Yttagran Island, Russia (see the Introduction for further details on this site). The bones are so light that even small groups of early native hunters could erect them. The bones of the larger seals, despite the time out the animal spends out of water, are also flimsy in comparison to those of terrestrial mammals. But the marine environment also has disadvantages for mammals. Heat loss in water is significantly higher than in air so the animals must have excellent body insulation, a high metabolic rate, or both. As volume (and therefore mass) increases with the cube of diameter, but surface area only with the square, large animals have a proportionately smaller surface area from which to lose heat than small ones. As a result, the smallest marine mammal is many thousands of times larger than the smallest terrestrial mammal. For insulation a layer of subcutaneous blubber is more efficient than fur, but as its insulating properties are, in part, dependent on thickness, good insulation requires a big body – another reason for marine mammals to be large. Photo 255b shows cut-up sections of a Bowhead killed by the Yuppiat people of Russia's northern Bering Sea coast and gives an indication of the thickness of the whale's blubber. The smaller marine mammals are those that rely, at least partially, on fur for insulation (Photo 255e is a Northern Fur Seal: the fur of this species was highly valued by fur trappers, so much so that the seals were almost hunted to extinction. The blubber of pinnipeds and cetaceans is of variable thickness and lipid content, its distribution optimising streamlining and the insulation of vital organs. Marine mammals also employ counter-current heat exchangers to minimise heat loss.

The other disadvantage of being a marine mammal is the breathing of air. Not only must the animals come to the surface to breathe – a procedure that might itself create problems for Arctic marine mammals because of the extent of ice cover – they must also store oxygen for relatively long periods if they are to feed successfully. The easy answer would appear to be large lungs – and large body size would appear to be just the thing to accommodate them. But storing a large supply of air has limitations. Large, air-filled lungs would act as buoyancy tanks making diving more difficult, and as pressure increases with depth the collapsing lungs would compress the air, with potentially lethal side effects. At high concentrations oxygen is poisonous, while nitrogen is a narcotic. Bubbles of nitrogen and oxygen forming in the blood as the animal surfaced and the gases decompressed would also give rise to the bends, a problem that can be fatal to human sub-aqua divers. Marine mammals dive to prodigious depths – Sperm Whales (Photo 255d) are known to dive to 3,000m (and to stay submerged for more than two hours) – and must therefore overcome this problem. The muscles of marine mammals are rich in myoglobin, which 'stores' blood (in much the same way as haemoglobin) and releases it

gradually during a dive. Marine mammal blood is also rich in haemoglobin, the oxygen storage potential of the two compounds reducing the need to store air when diving. Pinnipeds actually exhale before diving, effectively eliminating air storage and, therefore, the potential for the bends. But cetaceans inhale before diving. They possess networks of blood vessels known as a *rete mirabilia* (literally 'wonderful network'), in the chest cavity (and other areas), which, it is believed, may act as a sink for nitrogen as the animal surfaces. At the huge pressures of deep dives, the collapsing lungs of cetaceans also force air into the nasal passages, where nitrogen absorption into the bloodstream is not possible.

Pinnipeds

The Pinnipedia (the name means 'wing-footed') are divided into two superfamilies, the Phocoidea, which contains only one family, the Phocidae (true seals), and the Otarioidea, which contains two families, the Otariidae (eared seals – fur seals and sea lions) and the Odobenidae, which contains just one species, the Walrus. Pinnipeds origins are shrouded in mystery. Some authorities consider the two pinniped superfamilies represent two separate re-invasions of the sea, with an otter-like ancestor giving rise to the phocids and a bear-like ancestor the otarioids. However, molecular (DNA) research suggests a single re-invasion.→

True seals
Phocids live in both fresh and saltwater, though species such as the freshwater Baikal Seal (*Pusa sibirica*), the inland (but saltwater) Caspian Seal (*Pusa caspica*) and the Arctic-breeding marine Ringed Seal all evolved from the same ancestor. Apart from the monk seals, all phocids are polar or sub-polar, with all bar the two inland species of northern phocids being Arctic animals.

Ringed Seal *Pusa hispida*
The most numerous of all Arctic mammals, with a population estimated at over 5 million. Adults are dark grey or grey-brown above with a mosaic of pale grey rings (that give the seal its name), these smaller or absent on the head. The underparts are silver, silver-brown or grey-brown (Photos 257a-e, 'e' is a breathing/fishing hole, maintained open in the sea ice for easy access). The seals can dive to 90m, though depths to 40m are more common, and can stay submerged for 20 minutes, though 4–8 minutes is more usual. Seal pups are born in a chamber the female excavates above the sea ice, often in a pressure ridge. It is estimated that predation by bears and Arctic Foxes accounts for about 50% of each year's pups.

Breeds at the ice edge throughout the Arctic, with highest density in eastern Russia, the Bering Sea and North American Arctic. Also seen in three freshwater locations (Lake Ladoga in Russia, Lake Saimaa in Finland and Lake Nettilling in west Baffin Island). It is assumed that ancestral populations migrated to the lakes or were isolated by landscape changes.

Ribbon Seal *Phoca fasciata*
Adult males are dark brown or black with four distinct broad bands of pale grey and cream, one around the neck, two around the front flippers (these almost meeting on the breast) and a fourth around the lower abdomen (Photo 257f). Females have the same pattern, but the bands are much less distinct as the body colour is buff-brown. Breeds on pack ice far from land in the Bering Sea, southern Beaufort Sea and the Sea of Okhotsk.

Largha Seal *Phoca larga*
Adults are pale grey, grey-brown or mid-brown above, paler below with heavy, uniform dark brown and black spotting (Photos 257g and h: 'g' is a group of Largha Seals, and Slaty-backed Gulls on an islet off Kamchatka's east coast). Breeds in the North Pacific, the Sea of Okhotsk and the Bering Sea from Kamchatka to the western Alaska coast, and from the Aleutians to the Chukchi and Beaufort seas. The name is from the Tungus people of western Okhotsk. It was adopted because the North American name, Spotted Seal, though accurate, caused confusion with the Harbour Seal, which was sometimes called the Spotted Seal in Europe.

Harp Seal *Phoca groenlandica*
Adult males are silver-grey with dark grey or black upper muzzle, crown and cheeks, and a black 'harp' on the back and flanks. The 'harp' is actually seen on the flanks, the two harps linked across the back so that the overall pattern is more saddle–shaped. Adult females are patterned as the male, but the head and back/flank patch are paler. Harp Seal pups are pure white at birth, moulting to a silver-grey pelage with dark grey spotting and blotching at about 4 weeks. The annual slaughter of the helpless pups off Canada's eastern coast became a national and international issue some years ago, particularly as the killing, by clubbing, was seen as monstrous (though it was claimed that because of the fragile cranial bones, the method was actually instantaneous and so humane). Today the seals are still hunted, though the hunt is regulated, and carried out with less of the overt enthusiasm.

Found from the eastern Canadian Arctic islands east through Greenlandic waters/ North Atlantic to Russia's Laptev Sea, but with three distinct breeding populations near Newfoundland/Gulf of St Lawrence, Jan Mayen and the White Sea. After breeding the population migrates north, following the receding sea ice.

Harbour Seal *Phoca vitulina*
Adults form two colour morphs, dark morphs being silvery-grey with extensive dark grey, dark grey-brown or dark brown spotting dorsally, and on the crown and nape, while light morphs are buff, with fewer, and paler, markings. There are usually fewer ventral spots on either morph.

Harbour Seals are found in coastal waters of southern Greenland, Iceland, northern Scandinavia and Svalbard (north to Prinz Karls Forland), but are absent from northern Russia. They are, however, found in the northern Pacific and southern Bering Sea from Kamchatka to southern Alaska, and in the waters of eastern Canada, including Hudson Bay, northern Quebec and Labrador, and southern Baffin Island.

Bearded Seal *Erignathus barbatus*
Unusually, females are slightly larger than males. Adults are grey-brown or brown, darker above than below, and with some dark blotches. The head is small in comparison to the large, rotund body, making the seal look even fatter than it actually is. The seal has a profusion of long vibrissae that curl when dry, giving the moustached look that (more or less) explains the name.

Found throughout northern waters, and also in Hudson Bay and the Sea of Okhotsk.

→Pinnipeds are extremely well-adapted to the marine environment with spindle-shaped bodies, and limbs that have developed into flippers. However, there are significant differences between the two families. The phocids lack external ears (which would enhance drag), and have hind flippers closely akin to the tail flukes of whales in shape that provide the power for swimming. The front flippers are short and are held close to the body during swimming, though they can function as fins to aid steering. Insulation is by blubber, ancestral fur having been reduced to a sparse scattering of coarse hairs. On land the hind flippers are useless for locomotion, the seal using its front flippers to haul itself along, with progress being an ungainly wriggle.

One feature that all pinnipeds share is sensitive vibrissae, or whiskers. Studies have shown that these are sensitive to sound, which may be an advantage in avoiding predators. Studies with blindfolded seals have also shown that the seals are still able to catch fish, while studies in which the vibrissae have been removed indicate that the seals are then much less efficient at fishing. The suggestion is that the vibrissae can detect vibrations in the water such as those caused by the wake of a swimming fish. Vibrissae may also detect hydrodynamic changes caused by fixed features, and so help a seal to navigate in murky waters.→

---

a Male Harp Seal.
b Migrating Harp Seals off eastern Greenland.
c Mixed morph group of Harbour Seals.
d, e Light and dark morph Harbour Seals.
f, g Bearded Seals.

Hooded Seal   *Cystophora cristata*

Status: IUCN: Vulnerable (declining numbers due in part to animals being caught in the nets of commercial fisheries, but additional problems from reductions in sea ice coverage and oil spill/pollution).

Significant sexual dimorphism, males being almost twice the weight of females (though only about 25% longer). Males with weights exceeding 400kg have been recorded. The heaviest of the northern true seals, though not as long as the largest Grey Seals. Adults are silver-grey with extensive mottling of dark brown or black patches, these tending to be longer on the back and flanks. The hood of the name is an enlarged extension of the nasal cavity that forms a proboscis which, in males, hangs over the mouth, but is much less pronounced or absent in females. The hood can be inflated to form a large black cushion or blister that spreads from the forehead over the mouth. Males can also extrude and inflate the internasal septum membrane. This extrudes from one nostril, usually the left, as a red balloon. The inflation mechanisms of hood and balloon are dissimilar, the hood requiring closed nostrils, the balloon an open nostril. Consequently both hood and balloon cannot be inflated simultaneously. However a 'half-hood' and balloon can be inflated, the effect being grotesque. Although the hood and balloon are used in mating displays, they are also inflated if the seal is surprised by an observer (in anxiety or as a threat) and, occasionally, by resting seals, seemingly just for the fun of it. The hood develops in males from the age of 4 years.

Hooded Seals are found on the eastern seaboard of North America from Newfoundland to Lancaster Sound, but rarely west of Labrador's Cape Chidley or north into Smith Sound. Also found in the north-western Atlantic around Iceland's north coast, Jan Mayen, Svalbard and Bear Island, but rarely as far east as Franz Josef Land.

Grey Seal   *Halichoerus grypus*

The largest of the northern phocids. The species also exhibits the most striking sexual dimorphism, with the male up to three times larger than the female. Adult males are dark grey or grey-brown overall (though usually darker above than below) with light grey patches. Adult females are the reverse, being light grey with darker patches.

Subarctic rather than Arctic, Grey Seals are found on the Labrador coast, the southern and western coasts of Iceland, and northern Scandinavia eastwards to the White Sea.

→The smaller pinnipeds exhibit countershading, being darker above and paler below. The dorsal colour is also disrupted to break up the animal's outline when it hauls out. Ribbon and Harp Seals have striking patterns, which are more definite in males and which develop with age; these are likely to be related to courtship. Young seals are, in general, born with a covering of white fur (lanugo), a contrast to Antarctic seal pups, which are dark. The white fur is assumed to be camouflage against sea ice for animals born in the land of the Polar Bear, those pups that are not entirely white being pale or partially white, and having a disrupted pattern. The lanugo pelt of young seals is luxurious, compensating for the lack of blubber. The longer hairs of lanugo also trap air, adding extra insulation. Young seals also have brown fat which is metabolised as the first blubber layer is laid down.

Phocids moult their skin annually, but this is not equivalent to the moulting of fur-bearing mammals when losing or acquiring a winter pelt. For the seals such a change is unnecessary, the moult representing the replacement of potentially damaged skin.

a Hooded Seal.
b Male Hooded Seal with black nasal cavity proboscis half inflated and red internasal septum membrane inflated and extruded.
c-f Grey Seals. Photo 'c' clearly shows the lack of an external ear.

Eared Seals

In contrast to the True Seals, Eared Seals are able to rotate their hind flippers beneath their bodies. Using these and their long front flippers, the animals are then reasonably mobile and able to move surprisingly quickly. Travellers familiar with the slow-moving phocids are in for an unpleasant surprise if they stray too close to an eared seal. Not only does the animal accelerate and move quickly, it has an array of business-like teeth: being chased by an irate animal is akin to being pursued by a large dog, though thankfully the seal usually gives up more readily. Eared seals also differ in their insulation, relying, in part, on fur: the underfur of the fur seals is luxuriantly thick, a fact that led to their near-extinction due to overhunting. As a consequence of this different, and less effective, mode of insulation, fur seals and sea lions are chiefly animals of cool temperate waters. The Walrus, the sole member of the second otarioid family, is a true Arctic dweller: it has blubber insulation similar to the phocids though it shares the reversible hind flippers of the eared seals. Eared seals also differ in being gregarious, forming large, sometimes huge, colonies, whereas the phocids are more solitary.

Northern Fur Seal *Callorhinus ursinus*

Status: IUCN: Vulnerable (the population is declining, particularly in the Pribilofs, once the species' stronghold, though is slightly increasing at other breeding sites. Possible causes are food competition from commercial Alaska (or Walleye) Pollock (*Gadus chalcogrammus*) fishing, and possible change of habits of Orcas.

Extreme sexual dimorphism, with males being up to 5 times heavier than females and c.70% longer. Males may weigh 270kg. Adult males are rich dark brown, with the female being grey-brown above, pale chestnut-grey below. Males become darker with age and develop thickened necks and shoulders, and a mane of coarse hair. The hind flippers are very long, the largest of any member of the Otariidae. The seal's luxurious pelt is reflected in the scientific name, which derives from *kallos rhinos*, beautiful skin. It was for its pelt that the species was ruthlessly exploited in the 19th century when the original population on the Pribilof Islands was reduced from c.3million to c.300,000). The underfur of the seal has around 55,000 hairs/cm². only Sea Otters have a denser fur. So dense is the fur that water does not reach the skin even if the seal scratches itself under water.

Northern Fur Seals are found in the Sea of Okhotsk and across the Pacific from Japan to the Californian coast, but chiefly in the southern Bering Sea.

Steller's (or Northern) Sea Lion *Eumetopias jubatus*

Status: IUCN: Near threatened (the population of the western subspecies declined by 70% between 1977 and 2007 for unknown reasons. The decline has now halted and even reversed in some areas, though the population remains small, probably no more than 80,000 animals).

Though occasionally called the Northern Sea Lion, the more usual name remembers Georg Wilhelm Steller, the naturalist on Bering's expedition, who first described the species. Unlike other eared seals, Steller's Sea Lion, the largest of the eared seals and the most northerly sea lion, does have a layer of blubber, relying less on its fur for thermoregulation.

Significant sexual dimorphism, with males almost twice the weight of females (though only about 25% longer). Recorded male weights have exceeded 400kg. Adults vary from buff to red-brown, and are usually darker above than below. The flippers are dark grey or black. Males develop a thickened, muscular neck over which grows a mane of coarse hair.

Two subspecies are recognised, separated by genetic and subtle morphological differences, roughly divided by 144°W longitude. Nominate (Western Steller's) are the subarctic sea lions, being seen in the western Sea of Okhotsk and across the southern edge of the Bering Sea from Kamchatka through the Commander Islands to the Aleutians and southern Alaska (and also in the Kuril Islands). *E. j. monteriensis* (Loughlin's Steller) breeds along the North American coast south to central California.

a Northern Fur Seals on Pribilof. The size difference between males and females is clear.
b, c Male Northern Fur Seals, irate and relaxed about the presence of a photographer.
d Steller's Sea Lions relaxing on a buoy in Resurrection Bay, Alaska.
e Male Steller's Sea Lion.
f Group of female Steller's Sea Lions.

Walrus *Odobenus rosmarus*

Status: IUCN: Vulnerable (populations of both Atlantic and Pacific Walrus are in decline. Walrus require available shallow sea in which to feed and sea ice on which to haul out. For both populations the reduction in sea ice coverage, together with warming sea temperatures lead to stress which may be reflected in female condition and calf mortality. Although Atlantic Walrus are protected in Norway and Russia, they are not protected in Greenland where 'subsistence' continues. 'Subsistence' hunting of Pacific Walrus continues and is believed to be having an impact on the population. Land haul-out areas are also problematic because of induced stress, while sightseeing flights cause stampedes in which many young animals are crushed by large adults). CITES: Appendix III (Canada).

Walruses split from the eared seals about 20 million years ago. The fossil record suggests that walruses were once the dominant pinniped group, but gradually they declined, leaving just a single form today. Walruses are the largest of all Arctic pinnipeds, and second in size only to the elephant seals in world terms. The species shows extreme sexual dimorphism with males being c.50% heavier and c.20% longer than females. Some males may weigh over 2000kg.

The skin colour of adults varies with blood flow. In the water, or recently emerged, Walruses can be very pale grey or grey-brown. But when hauled out blood is pumped to the skin to aid cooling, and the animal becomes pink. Haul-outs with hundreds, perhaps thousands, of animals are a characteristic of the species and there are well-known haul out areas which have become famous (e.g. Pacific Walrus at Round Island, Alaska). However, as noted above, sightseers in boats and, particularly, aircraft can spook the animals. In late 2016 the inhabitants of Point Lay, on Alaska's north-west coast, formally asked visitors to stop arriving at the village to see the Walrus. The absence of sea ice caused the animals to haul out close to the village and in 2015 it was estimated that 35,000 arrived in the neighbourhood, the visitors overwhelming a village with no hotel or restaurant, and sightseeing aircraft causing mass stampedes.

Walrus skin is very thick (particularly around the neck, where it can be up to 4cm deep) and tough, and was used by the Inuit to cover summer and winter houses because of its durability. Walrus blubber can be up to 15cm thick, though on average it is only half that thickness. The upper canine teeth are massively extended to form protruding tusks. Atlantic and Pacific Walrus are different subspecies (Atlantic animals are nominate, Pacific are *O. r. divergens*: some authorities maintain the Walrus of the Laptev Sea are a third subspecies) and differ in the length and shape of their tusks. Male Atlantic tusks are (usually) straight and c.60cm, with those of the females to c.50cm. Pacific Walrus tusks are longer, those of males to 100cm, females to c.75cm, and they are also curved rather than straight (though there are exceptions). In general female tusks are circular in cross-section (male tusks are elliptical) and more slender. The Walrus uses its tusks to make or maintain holes in the ice, and to help hauling out of the water, the latter task giving the animal its scientific name – Odobenus from *odontes baino* – tooth-walker. The second part of the name derives from *ros maris* – sea rose, a reference to the colour of the animal and its maritime habitat. The common name is from the Scandinavian *hvalross* – whale horse.

Walrus feed by standing on their heads in shallow water and feeling for prey in the sediment with their highly sensitive vibrissae. The chief food is molluscs, the meat being extracted from the shell by suction. Walrus have a formidable ability to suck, the Inuit telling of animals coming up beneath swimming ducks and sucking them under. Walrus also eat young seals: the Inuit maintain they also occasionally kill Beluga. Adult Walrus have no enemies apart from humans. Polar Bears occasionally invade Walrus colonies, seeking to take a young animal. On land the Walrus is ponderous and no match for an agile bear, though bears ensure they stay well clear of adults as the tusks can inflict savage, potentially fatal, wounds. In water the tables are very definitely turned, the bears staying far away (see Photo 247d).

Walrus are found in north-west and north-east Greenland, Svalbard, Franz Josef Land, Novaya Zemlya, north-east Siberia, Wrangel Island, the west coast of Alaska, Baffin Island and the islands to the north of Hudson Bay, particularly near Igloolik.

a–d Atlantic Walrus.
e–g Pacific Walrus.

Cetaceans

As with pinnipeds, the whale shape is streamlined for energy-efficient locomotion. There is no discernible neck or shoulders: this means the head cannot move independently (the Beluga is an exception having unfused cervical vertebrae that allow the head to turn and nod). The hind limbs are vestigial and within the body so they do not interfere with the streamlining, the power for swimming being provided by a large tail comprising twin flukes. The tail is powered by huge back muscles and moves vertically, the flukes staying parallel to the water surface: whales therefore differ from fish, whose tails move side-to-side. The front limbs have become flippers primarily used for steering. Many species have evolved a dorsal fin to aid stability. Flippers and dorsal fin are the only protuberances. Body hair is minimised, insulation being entirely by subcutaneous blubber: Bowhead blubber can be 50cm thick.→

Baleen whales

Often called rorquals, a name of disputed origins, but probably from Scandinavia 'red throat' as the expanded furrows of the throat means the skin turns red as blood vessels are exposed, baleen whales represent only about 10% of all whale species, though in Arctic waters this rises to about 50% of species.

Blue Whale *Balaenoptera musculus*
Status: IUCN: Endangered (due to historical exploitation: now largely protected and increasing in numbers, but total population only 10,000-25,000). CITES: Appendix I.
The largest animal ever known to have existed, reaching 30m and 140t (Southern Ocean Blues may be 33m and 190t). Adults are long and narrow, blue-grey with occasional white blotching. Dorsal fin is small and falcate, and set far back. Notched tail, fluke trailing edge slightly concave Photos 283a and b). Feeds at 2-5km/h, but can reach speeds of up to 50km/h. Found in all oceans. Has been seen in the Bering Sea, in Baffin Bay and in the Barents Sea.

Fin Whale *Balaenoptera physalus*
Status: IUCN: Endangered (due to historical exploitation: now largely protected, but still hunted by Iceland (and others?): population and trend unknown). CITES: Appendix I.
The second largest whale, reaching 24m and 70t. Adults are long and narrow, slate-grey above, paler (even white) below. The dorsal fin is small and falcate, and set far back. Well-defined dorsal ridge from the fin to tail stock. Notched tail, fluke trailing edge shallowly convex, but upturned at the tips (Photos 283c and e). Similar speeds to Blue Whales, but may be faster in short bursts. Range as the Blue Whale, but tends to be more southerly.

Sei Whale *Balaenoptera borealis*
Status: IUCN: Endangered (due to historical exploitation: now largely protected: population and trend unknown). CITES: Appendix I.
The name, pronounced 'sigh' rather than 'say', derives from the Norwegian for pollock, once thought to be a principal prey. Females are up to 16m and 16t, and up to 40% larger than males by weight, but only 5% by length. Adults are mid- or dark grey above, paler below. The falcate dorsal fin is larger than those of Blue or Fin and closer to the head. Notched tail, fluke trailing edge straight with upturned tips (Photo 283d). Thought to be the fastest of the great whales. Found in all oceans. Tends to be more southerly than either Blue or Fin, rarely seen north of Jan Mayen, Labrador, south-east Greenland or the Aleutians.

Minke Whale *Balaenoptera acutorostrata*
Status: CITES: Appendix I (apart from West Greenland population, which is Appendix II).
Smallest Arctic baleen whale, up to 10.5m and 10t. Adults have flattened heads with a pointed snout, so from above the head forms a sharp V-shape. Dark grey above, diffusing into white patches on the flanks and ventral body. The dorsal fin is tall and falcate, and set far back on the body. Notched tail, fluke trailing edge shallowly concave (Photo 283f). Seen in the north Atlantic as far as Davis Strait to the west, occasionally reaching Svalbard and Barents Sea to the east. In north Pacific, found throughout Bering Sea and into Chukchi Sea.

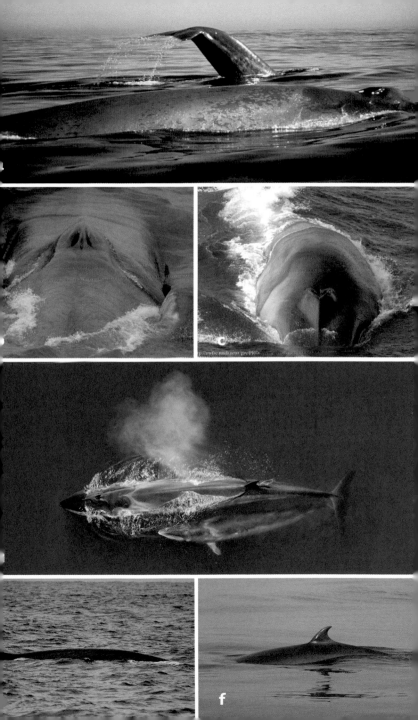

Humpback Whale *Megaptera novaeangliae*

Status: CITES: Appendix I.

Adults are up to 18m and 45t. Uniformly dark grey/black with vari...
underside. Front sections of the jaws have knoblike protuberance s...
stiff hair that is probably a sensory aid for detecting prey or water c...
285e). The falcate dorsal fin is small and mounted on a hump, us u...
front of the fin (Photo 285d: this is a family group). It is this hump...
common name. The pectoral fins are up to 5m long and have a dis...
pattern and knobbly leading edge. The pattern is highly individual...
individual whales. This is also true of the patterned flukes (Photo...
usually travelling at 2-6km/h, but capable of bursts to 30km/h. F...
285a is actually a double breach, the furthest whale having alread...
oceans. Seen in the Atlantic as far north as Svalbard, but rarely e a...
Pacific occurs in the Bering Sea. Photo 285g is of a Humpback bl...

Grey (Gray) Whale *Eschrichtius robustus*

Status: IUCN: Overall Least Concern, though the population of th...
Okhotsk is Critically Endangered. CITES: Appendix I.

Up to 15m and 35t. Grey Whales possess characteristics of both...
Adults are mottled dark grey, light grey and white, and invariab l y
the head and forward part of the back (Photo 285c). There is no...
triangular dorsal hump, from which a series of small bumps run...
the tail stock. Notched tail, the flukes broad, with straight (thou...
upturned tips. Because of their feeding method, ploughing thro...
whales stir up clouds of silt and food particles, so feeding whale...
of the accompanying flocks of gulls and other seabirds. Found i...
as the Chukchi and Beaufort seas. There was formerly an Atlant...
feeding off Iceland and Greenland and migrating as far south as...
whales became extinct in the 18th century, probably due to over...
Grey Whales move between the Bering Sea and the lagoons of...
California, where birthing and mating take place, and back eac...
or more. Newborn calves accompany mothers during the swi...
the whales (evolution? or only individuals?) have 'acquired' a st...
Inuits, who hunt them, find them inedible (as do their dogs).

→Whales are considered to have evolved from a common a...
branches emerging to form the two modern-day suborders,...
Although the two groups have many common features,...
methods and, consequently, in the structure of the head. Toot...
and squid. In general their jaws are extended into a beak-li k...
beaked whales, the Sperm Whale being the major exceptio n...
for grasping prey (or tearing at it in the case of the Orca). Th...
'melon' within which is a wax-like substance that is the basis...
find prey in the deep, dark waters in which the whales tend to...
huge, the head accounting for 25-30% of total length.

The structure of the head is very different in the baleen...
are extended and widened, the upper bones forming the rost...
hang. Often called whalebone, baleen plates are actually ke r...
from the jaw bone. The plates are smooth, but the inner ed...
overlapping to form a sieve that captures food as the whale...
presses against the baleen engulfed water is squeezed out th r...
then being swallowed. Although this action is common, bale...
strategies. Some swim slowly forward, their sieves extracti...
huge gulps of water. Sievers include Right and Bowhead...
allow space for large baleen plates. Gulpers include Blue a...
furrows of skin on the lower jaw, which allows the mouth to...
water at each gulp. Sei Whales feed with a combination of s...
differs in sieving bottom sediments.→

Walrus *Odobenus rosmarus*

Status: IUCN: Vulnerable (populations of both Atlantic and Pacific Walrus are in decline. Walrus require available shallow sea in which to feed and sea ice on which to haul out. For both populations the reduction in sea ice coverage, together with warming sea temperatures lead to stress which may be reflected in female condition and calf mortality. Although Atlantic Walrus are protected in Norway and Russia, they are not protected in Greenland where 'subsistence' continues. 'Subsistence' hunting of Pacific Walrus continues and is believed to be having an impact on the population. Land haul-out areas are also problematic because of induced stress, while sightseeing flights cause stampedes in which many young animals are crushed by large adults). CITES: Appendix III (Canada).

Walruses split from the eared seals about 20 million years ago. The fossil record suggests that walruses were once the dominant pinniped group, but gradually they declined, leaving just a single form today. Walruses are the largest of all Arctic pinnipeds, and second in size only to the elephant seals in world terms. The species shows extreme sexual dimorphism with males being c.50% heavier and c.20% longer than females. Some males may weigh over 2000kg.

The skin colour of adults varies with blood flow. In the water, or recently emerged, Walruses can be very pale grey or grey-brown. But when hauled out blood is pumped to the skin to aid cooling, and the animal becomes pink. Haul-outs with hundreds, perhaps thousands, of animals are a characteristic of the species and there are well-known haul out areas which have become famous (e.g. Pacific Walrus at Round Island, Alaska). However, as noted above, sightseers in boats and, particularly, aircraft can spook the animals. In late 2016 the inhabitants of Point Lay, on Alaska's north-west coast, formally asked visitors to stop arriving at the village to see the Walrus. The absence of sea ice caused the animals to haul out close to the village and in 2015 it was estimated that 35,000 arrived in the neighbourhood, the visitors overwhelming a village with no hotel or restaurant, and sightseeing aircraft causing mass stampedes.

Walrus skin is very thick (particularly around the neck, where it can be up to 4cm deep) and tough, and was used by the Inuit to cover summer and winter houses because of its durability. Walrus blubber can be up to 15cm thick, though on average it is only half that thickness. The upper canine teeth are massively extended to form protruding tusks. Atlantic and Pacific Walrus are different subspecies (Atlantic animals are nominate, Pacific are *O. r. divergens*: some authorities maintain the Walrus of the Laptev Sea are a third subspecies) and differ in the length and shape of their tusks. Male Atlantic tusks are (usually) straight and c.60cm, with those of the females to c.50cm. Pacific Walrus tusks are longer, those of males to 100cm, females to c.75cm, and they are also curved rather than straight (though there are exceptions). In general female tusks are circular in cross-section (male tusks are elliptical) and more slender. The Walrus uses its tusks to make or maintain holes in the ice, and to help hauling out of the water, the latter task giving the animal its scientific name – Odobenus from *odontes baino* – tooth-walker. The second part of the name derives from *ros maris* – sea rose, a reference to the colour of the animal and its maritime habitat. The common name is from the Scandinavian *hvalross* – whale horse.

Walrus feed by standing on their heads in shallow water and feeling for prey in the sediment with their highly sensitive vibrissae. The chief food is molluscs, the meat being extracted from the shell by suction. Walrus have a formidable ability to suck, the Inuit telling of animals coming up beneath swimming ducks and sucking them under. Walrus also eat young seals: the Inuit maintain they also occasionally kill Beluga. Adult Walrus have no enemies apart from humans. Polar Bears occasionally invade Walrus colonies, seeking to take a young animal. On land the Walrus is ponderous and no match for an agile bear, though bears ensure they stay well clear of adults as the tusks can inflict savage, potentially fatal, wounds. In water the tables are very definitely turned, the bears staying far away (see Photo 247d).

Walrus are found in north-west and north-east Greenland, Svalbard, Franz Josef Land, Novaya Zemlya, north-east Siberia, Wrangel Island, the west coast of Alaska, Baffin Island and the islands to the north of Hudson Bay, particularly near Igloolik.

a-d Atlantic Walrus.
e-g Pacific Walrus.

Cetaceans

As with pinnipeds, the whale shape is streamlined for energy-efficient locomotion. There is no discernible neck or shoulders: this means the head cannot move independently (the Beluga is an exception having unfused cervical vertebrae that allow the head to turn and nod). The hind limbs are vestigial and within the body so they do not interfere with the streamlining, the power for swimming being provided by a large tail comprising twin flukes. The tail is powered by huge back muscles and moves vertically, the flukes staying parallel to the water surface: whales therefore differ from fish, whose tails move side-to-side. The front limbs have become flippers primarily used for steering. Many species have evolved a dorsal fin to aid stability. Flippers and dorsal fin are the only protuberances. Body hair is minimised, insulation being entirely by sub-cutaneous blubber: Bowhead blubber can be 50cm thick.→

Baleen whales

Often called rorquals, a name of disputed origins, but probably from Scandinavia 'red throat' as the expanded furrows of the throat means the skin turns red as blood vessels are exposed, baleen whales represent only about 10% of all whale species, though in Arctic waters this rises to about 50% of species.

Blue Whale *Balaenoptera musculus*
Status: IUCN: Endangered (due to historical exploitation: now largely protected and increasing in numbers, but total population only 10,000-25,000). CITES: Appendix I.
The largest animal ever known to have existed, reaching 30m and 140t (Southern Ocean Blues may be 33m and 190t). Adults are long and narrow, blue-grey with occasional white blotching. Dorsal fin is small and falcate, and set far back. Notched tail, fluke trailing edge slightly concave Photos 283a and b). Feeds at 2-5km/h, but can reach speeds of up to 50km/h. Found in all oceans. Has been seen in the Bering Sea, in Baffin Bay and in the Barents Sea.

Fin Whale *Balaenoptera physalus*
Status: IUCN: Endangered (due to historical exploitation: now largely protected, but still hunted by Iceland (and others?): population and trend unknown). CITES: Appendix I.
The second largest whale, reaching 24m and 70t. Adults are long and narrow, slate-grey above, paler (even white) below. The dorsal fin is small and falcate, and set far back. Well-defined dorsal ridge from the fin to tail stock. Notched tail, fluke trailing edge shallowly convex, but upturned at the tips (Photos 283c and e). Similar speeds to Blue Whales, but may be faster in short bursts. Range as the Blue Whale, but tends to be more southerly.

Sei Whale *Balaenoptera borealis*
Status: IUCN: Endangered (due to historical exploitation: now largely protected: population and trend unknown). CITES: Appendix I.
The name, pronounced 'sigh' rather than 'say', derives from the Norwegian for pollock, once thought to be a principal prey. Females are up to 16m and 16t, and up to 40% larger than males by weight, but only 5% by length. Adults are mid- or dark grey above, paler below. The falcate dorsal fin is larger than those of Blue or Fin and closer to the head. Notched tail, fluke trailing edge straight with upturned tips. Thought to be the fastest of the great whales (Photo 283d). Found in all oceans. Tends to be more southerly than either Blue or Fin, rarely seen north of Jan Mayen, Labrador, south-east Greenland or the Aleutians.

Minke Whale *Balaenoptera acutorostrata*
Status: CITES: Appendix I (apart from West Greenland population, which is Appendix II).
Smallest Arctic baleen whale, up to 10.5m and 10t. Adults have flattened heads with a pointed snout, so from above the head forms a sharp V-shape. Dark grey above, diffusing into white patches on the flanks and ventral body. The dorsal fin is tall and falcate, and set far back on the body. Notched tail, fluke trailing edge shallowly concave (Photo 283f). Seen in the north Atlantic as far as Davis Strait to the west, occasionally reaching Svalbard and Barents Sea to the east. In north Pacific, found throughout Bering Sea and into Chukchi Sea.

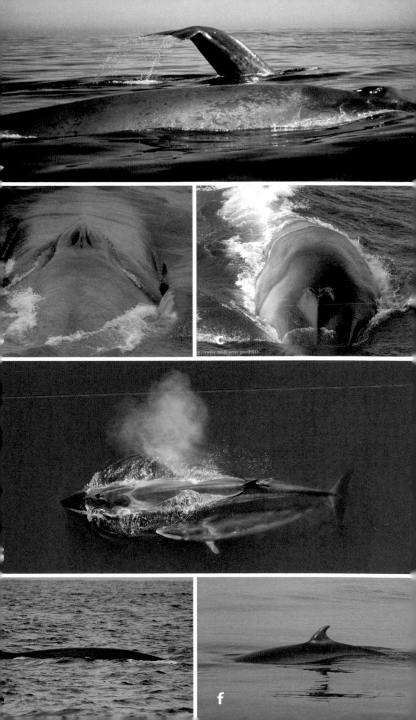

Humpback Whale *Megaptera novaeangliae*

Status: CITES: Appendix I.

Adults are up to 18m and 45t. Uniformly dark grey/black with variable white patches on the underside. Front sections of the jaws have knoblike protuberances, each of which encloses a stiff hair that is probably a sensory aid for detecting prey or water current movements (Photo 285e). The falcate dorsal fin is small and mounted on a hump, usually more easily visible in front of the fin (Photo 285d: this is a family group). It is this hump that gives the species its common name. The pectoral fins are up to 5m long and have a distinctive black-and-white pattern and knobbly leading edge. The pattern is highly individual and is used to identify individual whales. This is also true of the patterned flukes (Photo 285f). Slow swimmers, usually travelling at 2–6km/h, but capable of bursts to 30km/h. Famed for breaching (Photo 285a is actually a double breach, the furthest whale having already 'landed'). Found in all oceans. Seen in the Atlantic as far north as Svalbard, but rarely east of the Barents Sea. In the Pacific occurs in the Bering Sea. Photo 285g is of a Humpback blow.

Grey (Gray) Whale *Eschrichtius robustus*

Status: IUCN: Overall Least Concern, though the population of the north-west Pacific and Sea of Okhotsk is Critically Endangered. CITES: Appendix I.

Up to 15m and 35t. Grey Whales possess characteristics of both rorquals and Right Whales. Adults are mottled dark grey, light grey and white, and invariably have barnacle clusters on the head and forward part of the back (Photo 285c). There is no dorsal fin, merely a small, triangular dorsal hump, from which a series of small bumps runs along the dorsal ridge to the tail stock. Notched tail, the flukes broad, with straight (though ragged) trailing edges and upturned tips. Because of their feeding method, ploughing through bottom sediments, the whales stir up clouds of silt and food particles, so feeding whales are often noticeable because of the accompanying flocks of gulls and other seabirds. Found in the North Pacific, as far north as the Chukchi and Beaufort seas. There was formerly an Atlantic population, the animals feeding off Iceland and Greenland and migrating as far south as the Bay of Biscay, but the whales became extinct in the 18th century, probably due to overhunting by Basque whalers. Grey Whales move between the Bering Sea and the lagoons of Baja California and the Gulf of California, where birthing and mating take place, and back each year, a distance of 12,000km or more. Newborn calves accompany mothers during the swim north (Photo 285b). Recently the whales (evolution? or only individuals?) have 'acquired' a stink and foul taste so that Yuppik Inuits, who hunt them, find them inedible (as do their dogs).

→Whales are considered to have evolved from a common ancestor in the Eocene, with two branches emerging to form the two modern-day suborders, the toothed and baleen whales. Although the two groups have many common features, they differ markedly in feeding methods and, consequently, in the structure of the head. Toothed whales feed primarily on fish and squid. In general their jaws are extended into a beak-like snout (most pronounced in the beaked whales, the Sperm Whale being the major exception). The jaws have an array of teeth for grasping prey (or tearing at it in the case of Orca). The forehead is rounded, forming a 'melon' within which is a wax-like substance that is the basis of an echolocation system used to find prey in the deep, dark waters in which the whales tend to feed. The Sperm Whale melon is huge, the head accounting for 25–30% of total length.

The structure of the head is very different in the baleen whales. The cranial and jaw bones are extended and widened, the upper bones forming the rostrum, from which the baleen plates hang. Often called whalebone, baleen plates are actually keratinous (hair-like) plates emerging from the jaw bone. The plates are smooth, but the inner edges abrade to form 'bristles', these overlapping to form a sieve that captures food as the whale swims forward. When the tongue presses against the baleen engulfed water is squeezed out through the sieves, trapped prey items then being swallowed. Although this action is common, baleen whales have different engulfing strategies. Some swim slowly forward, their sieves extracting food continuously, others take huge gulps of water. Sievers include Right and Bowhead Whales which have huge heads to allow space for large baleen plates. Gulpers include Blue and Humpback: they have pleats or furrows of skin on the lower jaw, which allows the mouth to expand and engulf vast quantities of water at each gulp. Sei Whales feed with a combination of sieving and gulping: the Grey Whale differs in sieving bottom sediments.→

Right whales

Right whales acquired their name from the early whalers – these were the right (i.e. correct) whales to kill: they were slow and so could be easily overhauled by a rowed boat; they were passive, and so did not turn every killing into a battle in which the whalers were at risk; they floated when dead, making them easier to transport to ships or shore; and they yielded huge amounts of baleen and oil. This lethal combination led to the animals being hunted almost to extinction. Despite full protection (though the Bowhead is still hunted by native peoples in both Alaska and Siberia) the populations do not appear to be recovering well, possibly due to inbreeding. There are anatomical differences between Right whales and rorquals: Right whales have an arched nostrum, forming bow-shaped rather than the straight mouth of the rorquals, and so have longer baleen plates, and have no throat pleats, feed by skimming rather than gulping. Also, there is no dorsal fin.

Northern Right Whale *Eubalaena glacialis*

Status: IUCN: Critically Endangered (now fully protected, but may be below species viability threshold. The eastern Atlantic population may extinct. The western Atlantic population is 300-350, but perhaps with only 250 mature animals. In June-August 2017 10 carcasses were discovered floating near/beached on eastern Canada, representing perhaps 2-3% of the population and double the annual calving rate. Deaths were unexplained, but possibly due to ship strikes or net entanglements. Pacific population is classified as Endangered: numbers unknown, but probably less than 1000 animals). CITES: Appendix I.

One of the world's rarest cetaceans, up to 18m and 90t. Some authorities believe Pacific animals are a separate species (*E. japonica*) not a subspecies. Photo 287a, female and calf: Photos 287b and c are west Atlantic Right Whales. Adults are large and rotund; black with variable ventral white patches. The head has a number of callosities (usually white, but occasionally yellow or pink), which are also seen in foetal whales: their function is unknown. Blubber is up to 60cm thick and contributes 40% of the total weight. Deeply notched tail, flukes have concave trailing edge. Rarely exceeds 10km/h. Found between Iceland and Norway (now extinct?), between Labrador and Maine, and from Japan to Kamchatka and in the Sea of Okhotsk. Rare near the Aleutians and southern Alaska.

Bowhead Whale *Balaena mysticetus*

Status: CITES: Appendix I.

The most Arctic of all whales, up to 18.5m and 80t. Massive head makes up 35-40% of total length and bears the huge bow-shaped mouth. The adult is black with a white lower lip, marked by black spots (Photo 287f: Photo 287d is a sleeping calf at the ice edge). The back is often marked with white, these marks thought to be scarring acquired when the whales break through the sea ice to breathe. Bowheads can break through ice up to 60cm thick. There is a prominent triangular bump in front of the blowholes, and a depression behind them. Deeply notched tail, the flukes have a shallow concave trailing edge. Swims slowly at about 6km/h. Bowheads taken by native hunters (e.g. Photo 287e, Chukotka) have been found to have ancient harpoon heads embedded in them, these dated to well over 100 years old. Studies suggest Bowheads may live up to 200 years, and were considered to rival Giant Tortoises (*Geochelone* sp.) as the longest-lived animals but very recent studies (late 2016) suggest that Greenland Sharks (*Somniosus microcephalus*) may live for 400 years.

→In toothed whales the two nasal passages combine to form a single, crescent-shaped blowhole. In some species it seems that only one passage is used for breathing, the second being part of the echolocation system. In baleen whales the nasal passage form a blowhole of two parallel slits. When the whale exhales, water trapped in folds around the blowhole is expelled, forming the characteristic 'blow': the blow pattern is a useful guide to species.

Cetacean eyes are relatively small, sight being much less useful in water. However, some whales 'spy hop', raising their heads out of the water, apparently to view the local area. Small eyes then act as pinhole cameras, allowing greater depth of focus, and may assist the whales – which mostly stay close to shore – to locate land features. Spy-hopping is one of many distinctive whale behaviours. Others include lob-tailing (water slapping with the tail), fin-waving (which may involve water slapping), and breaching. These are all forms of communication, but are more common in some species than others. Humpbacks are famous for breaching, but surprisingly the huge, slow Bowhead and Northern Right whales also do it.

Toothed whales

Species of the suborder Odontoceti, the toothed whales, make up the majority of cetaceans. Most are surface feeders and show similar countershading to that of pinnipeds.

Beluga *Delphinapterus leucas*

Status: IUCN: Near threatened (due largely to both commercial and subsistence hunting, but also climate change and industrialisation. The population and trend are poorly known, particularly in Russia). CITES: Appendix II.

One of the species in the family Monodontidae, both of which are true Arctic dwellers. Adults, which are up 5m and 1.5t, are entirely creamy-white (the name derives from the Russian for 'white'). There is no dorsal fin, but a small triangular ridge is visible in many individuals. The head is broad with a distinctive 'melon' that becomes larger with age. Calves are uniformly grey, becoming white only when they about 6 years old. Swims at 5-15km/h. Unusually for a cetacean can move their heads, being able to both turn and nod the head. Equally unusual is the fact that Belugas moult annually, choosing river mouths where the warmth and low salinity of the mixed fresh and sea waters aid the process, and shallow water allows the animals to rub the old skin free on the river bed. As well as styling itself the 'Polar Bear Capital of the World' because of the annual congregation of bears, Churchill, Manitoba, at the south-western corner of Hudson Bay, also styles itself the 'Beluga Capital of the World' for the summer congregation of whales in the Churchill River.

Belugas are found near Svalbard and east from there along the Russian Arctic coast to the Chukchi Sea. Also found in the Bering Sea and the Arctic waters of North America to eastern Greenland, and in Hudson Bay. However, the animals are very rare in the Greenland Sea (and perhaps absent altogether).

Narwhal *Monodon monoceros*

Status: IUCN: Near threatened (due to hunting, climate change and industrialisation: the population is believed to be about 80,000). CITES: Appendix II.

Adults, which are up to 5.5m, excluding the tusk, and 1.6t, are mottled blue-grey or dark grey and white, the mottling usually more extensive on the upperparts. There is no dorsal fin, and the dorsal ridge is marked only by a dark line. Calves are uniformly mid-grey, becoming darker and more mottled as they mature. Older adults become paler, some being almost white. The flukes are convex in males, less so, or even straight, in females. Narwhals have only two teeth, both in the upper jaw. In males the left tooth pushes through the lip to form a tusk. In some males the right tooth also erupts, though these double-tusked narwhals are very rare. The tusk begins to grow when the animal is 2-3 years old (occasionally at one year) and grows continuously. Very old males may have tusks of 3m weighing 10kg. In females the teeth often do not erupt during the entire life of the animal, though tusked females have been seen. The Narwhal's tusk is claimed to be the source of tales of the Unicorn. This cannot be conclusively proved, but it does appear that the true source of the tusks was suppressed to promote the Unicorn legend (and hence the price of the tusks). Despite occasional nonsense written on the subject, the tusk is not used to skewer fish. The tusk is a secondary sexual characteristic. It is sometimes used as a weapon: scarred males and broken tusks are seen. The tusk is also sometimes laid across the back of another animal in what appears to be a gentle, tactile gesture. The whale's curious name is from the Scandinavian *nár hvalr*, corpse whale, because the skin colour looks like that of a dead man. Narwhals swim at 5-15km/h. Narwhal skin (*muktuq*) is considered a great delicacy by the Inuit and is eaten as soon as a hunted animal is landed. It tastes, vaguely, of hazelnut.

Narwhals are a true Arctic dweller, being found from west Greenland to the New Siberian Islands (though they are rare throughout that area). They are more common in the Canadian Arctic, from Banks Island to east Greenland.

a Hunting Belugas always attract seabirds, in this case American Herring Gulls and one Thayer's Gull.
b, c Adult, left, and juvenile, right, Beluga.
d Narwhal in a pack-ice lead.
e Narwhal tusking.

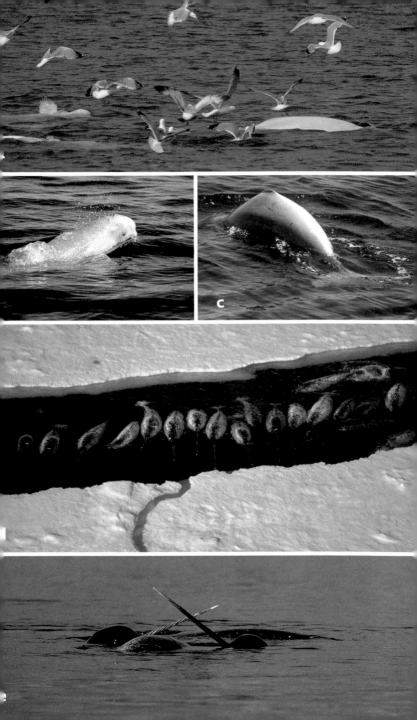

Sperm Whale *Physeter macrocephalus*

Status: IUCN: Vulnerable (overhunted historically and population rise following hunting ban seems very slow: population and trend poorly understood). CITES: Appendix I.

The largest toothed whale (males to 18.5m and 24t), with a legendary place in both the history of whaling, because of the ferocity of some whales towards their pursuers, and in literature as a result of Herman Melville's classic novel *Moby Dick*. Exhibits the most extreme sexual dimorphism of any cetacean, with males weighing up to three times more than the female. Adults are dark grey, with occasional white patches. Surfacing whales often appear more grey-brown. Moby Dick was, of course, white, and white Sperm Whales are known, these being truly white rather than cases of albinism. The lower jaw is long and narrow, and holds 18–25 conical teeth in each half. These teeth grow to around 25cm and fit into sockets in the upper jaw when the mouth is closed. The upper jaw has no teeth or just a few rudimentary ones. The dorsal fin is small, thick and rounded, and behind it there is a series of bumps along the dorsal ridge towards the tail stock. Notched tail, the flukes with a straight trailing edge. Sperm Whales usually swim at 4–6km/h, but can reach 25km/h.

Sperm Whales are found in all oceans to the ice edge, but is rare in the far north.

North Atlantic Bottlenose Whale *Hyperoodon ampullatus*

Status: IUCN: Data deficient (overhunted historically and at risk due to climate change: population and trend poorly understood). CITES: Appendix I.

Adults, up to 9.5m and 7.5t, vary from dark brown through grey-brown to greenish brown, usually paler ventrally with cream or cream–buff blotches. The melon is very prominent and is cream from crown to forehead. The dorsal fin is small and triangular, with a falcate trailing edge. The tail is not notched, the fluke trailing edge straight or shallowly concave with sharply upturned tips.

Found in the north Atlantic to the ice edge, but rare in the Barents Sea and Hudson Strait.

Orca (Killer Whale) *Orcinus orca*

Status: IUCN Data deficient (at risk from pollutants and killing by commercial fishermen. The population and trend is poorly understood). CITES: Appendix II.

The Orca is the largest member of the dolphin family, males up to 9m and 4t, females about a third smaller. Adults are black above, white below, with a pale grey saddle behind the prominent dorsal fin. The male dorsal fin is tall and triangular (up to 2m high), in females it is short and falcate. The tail flukes form a shallow V. Orcas can swim at up to 60km/h, but usually swim at 5–10km/h. They are social animals, occasionally forming large pods.

Orcas are found in all oceans. In the north they are seen to the ice edge in both the North Atlantic and North Pacific.

White-beaked Dolphin *Lagenorhynchus albirostris*

Status: CITES: Appendix II.

Adults (up to 2.8m and, exceptionally, 350kg) are highly variable, but usually the back is black to the falcate dorsal fin (taller in the male than the female), then pale grey or white to the black tail stock. The flanks are striped in shades of grey with a black patch forward of, and below, the dorsal fin. The ventral side is white. Notched tail, the flukes having concave trailing edges.

Found in the North Atlantic. It is the most northerly of the smaller dolphins, reaching the southern shores of Svalbard and the Barents Sea, though more southerly in the colder western Atlantic where it is rarely seen north of Labrador or south-west Greenland. Despite the northerly range the animals are poor ice travellers and many die after becoming entrapped in the pack ice.

a, b Sperm Whales.
c North Atlantic Bottlenose Whales.
d-f Orcas.
g, h White-beaked Dolphins.

In this final section on toothed whales we consider species which are not truly Arctic, but which may, as the oceans warm, be seen in Arctic waters. Such species include the Atlantic White-sided Dolphin (*Lagenorhynchus acutus*), the Pacific White-sided Dolphin (*Lagenorhynchus obliquidens*), the Common Bottlenose Dolphin (*Tursiops truncatus*) and Sowerby's (or the North Sea) Beaked Whale (*Mesoplodon bidens*).

Long-finned Pilot Whale *Globicephala melas*

Status: IUCN Data deficient (hunted in Faroes and Greenland: some evidence of population decline, but population and trend poorly understood). CITES: Appendix II.

Second largest dolphin after the Orca, up to 7m and 3.5t. Adults are dark grey with several paler ventral patches and paler banding on the head. Has a bulbous head (the scientific name translates as 'globe headed, black'– Photo 293c). Short falcate dorsal fin with a long base, set relatively far forward on the body. Notched tail, flukes having a straight or slightly concave trailing edge. Found in the North Atlantic north to Iceland and southern Greenland.

Baird's Beaked Whale *Berardius bairdii*

Status: IUCN: Data deficient (hunted in Japan and, in small numbers in Canada, Russia and USA: also at risk from climate change and loud anthropogenic noises: population and trend poorly understood). CITES: Appendix I.

Adults are up to 13m and 11t, dark grey/grey-brown with numerous white ventral patches. Lower jaw longer than the upper and the two anterior teeth of the four in the jaw are visible when the mouth is closed. The dorsal fin is small with a straight or falcate trailing edge. Notched tail, the flukes have a shallow concave trailing edge (Photo 293a). Found north of a line from Japan to Baja California, as far north as the Sea of Okhotsk and the Pribilof Islands.

In July 2016 a dead whale washed up in the Pribilofs. Examination confirmed tales of *karasu* (raven) by Japanese fishermen, a new species – a rare, dark whale similar to Baird's.

Stejneger's Beaked Whale *Mesoplodon stejnegeri*

Status: IUCN: Data deficient (at risk from entanglement in nets, from climate change and loud anthropogenic noises: population and trend poorly understood). CITES: Appendix II.

Known only from a single skull described in 1885, and two dead whales washed ashore in Alaska, and virtually unseen until 1994 when a group of four stranded on Adak Island in the Aleutians. Almost nothing is known of its biology. Up to 6m and 1.2t. Males are dark grey overall, females are paler ventrally. Has two tusk-like teeth in the lower jaw, set about 20cm back from the beak tip and protruding well above the upper jaw (Photo 293d, Marine World, Fukuoka, Japan). Small, triangular Dorsal fin with a falcate trailing edge. Tail not notched, the flukes have a straight trailing edge, but upturned at the extremities. Appears confined to a curved band of the north Pacific from Japan to northern California, and in the Bering Sea north to southern Chukotka, but much more southerly on the North American side.

Dall's Porpoise *Phocoenoides dalli*

Status: CITES: Appendix II.

Adults, up to 2.4m and 210kg, are black with large white patches on the flanks stretching from just in front of the dorsal fin almost to the tail stock, and continuous ventrally. There are also variable white patches on the dorsal fin, and on the trailing edges of both flippers and flukes. Adult males have a dorsal hump forward of the dorsal fin. The dorsal fin is triangular with a hooked top. The flukes are notched (Photo 293b). Found in the Sea of Okhotsk and from Kamchatka to Alaska, though rarely far north in the Bering Sea.

Common (Harbour) Porpoise *Phocoena phocoena*

Status: IUCN: Least concern (though population and trend poorly understood, but surveys in small areas suggest global population of 700,000: was considered Vulnerable until 2008). CITES: Appendix II.

Adults, up to 1.9m and exceptionally to 95kg, are dark grey or black above, diffusing into pale grey on the flanks and paler grey or white below. The dorsal fin is short and triangular, with a concave trailing edge. The tail has a distinct notch. (Photos 293d and e). Occurs from Iceland east to Novaya Zemlya, around southern Greenland and off Labrador coast, and from southern Kamchatka across the Aleutians to south-west and southern Alaska.

Sea Ice, the Sea and Freshwater

The Arctic Ocean can paradoxically be described as among both the least and the most productive seas on Earth. The central ocean, where thick multi-year ice restricts sunlight transmission to the open water beneath, is an area of very low productivity, but the shallow seas above the extensive continental shelves that surround the ocean are, seasonally, highly productive. The reduced sunlight of winter, coupled with seasonal sea ice, restricts the growth of the photosynthetic organisms, but the sunlight of the Arctic summer, coupled with the nutrients that flow in from the huge continental rivers, increases productivity dramatically.

Even though the sea ice restricts light transmission and so limits sub-ice productivity, the ice itself is not devoid of life. Surface meltwater pools support micro-organisms deposited by overtopping waves, river water inflow or the feet of seabirds, flourish. Other organisms live on the lower surface of the ice. Within the ice lives phytoplankton – diatoms and other types of algae. Diatoms are single-celled organisms that reproduce by division: more than 200 species have been identified so far living within the Arctic ice. They live in the brine channels of the ice matrix, a home that demands not only an ability to withstand very low temperatures, but one that also requires adaptations to cope with the osmotic pressures created by high salinity. The reproduction of the diatoms depends on local temperature. Although they can survive at low temperatures their growth rate is slow: a diatom that might divide every day at 0°C, would perhaps take three days at -4°C, and 50 days or more at -8°C.

Diatoms stain the ice brown (Photo 295a): as this darker colour absorbs more heat, the local ice temperature increases, causing melting, and the multiplying diatoms spread into the honeycomb structure this creates. Larger organisms feed on the diatoms and other phytoplankton (these also live within the ice matrix – the juvenile stages of many crustaceans are found within or on the underside of the ice). These amphipods, copepods and euphausiids are themselves part of a food chain that includes fish, marine mammals and seabirds at higher trophic levels. Phytoplankton also live on the underside of the ice; one diatom, *Melosira arctica*, forms filaments and even sheets that can grow up to 15m long. These filamentary structures act as nets, trapping nutrients from the water (Photo 295b).

Where sea ice melts, photosynthesis occurs at all depths to which light penetrates and photosynthetic phytoplankton make a living: during the winter months these enter a state comparable to hibernation. The most regularly encountered marine photosynthetic organisms are seaweeds (Photo 295c). These grow surprisingly far north: some intertidal species can survive temperatures as low as -60°C. Some species can actually begin to grow during the late stages of the Arctic winter, i.e. in darkness, using starches stored during the previous summer. Eelgrass (*Zostera marina*), a flowering plant that thrives in the shallow water of estuaries and tidal lagoons, also provides a rich microhabitat (Photo 295d): crustaceans and small fish live on and within the plants, while geese and other wildfowl feed on it. When the ice melts and sunlight increases, the phytoplankton blooms as do the herbivorous zooplankton that feeds on them. The most abundant of these are the Calanus copepods (particularly *Calanus glacialis* and *C. finmarchicus*) and amphipods (particularly *Apherusa glacialis* and *Gammarus wilkitzkii*). Most adult crustaceans do not penetrate the ice matrix, due in part to their inability to squeeze into the brine channels, but also because they are unable to cope with the high salinity. Copepods are the crustaceans most likely to be found within the ice. Those that do not enter the ice feed on diatoms that fall from the matrix.

The herbivorous zooplankton sustains an array of small fish and on up the food chain. Over 100 species of fish have been identified in Arctic waters. Of these, two of the most important are the Arctic Cod (*Boreogadus saida*: Photo 297a) and Glacial Cod (*Arctogadus glacialis*) smaller cousins of southern cod. Arctic Cod use cracks within the sea ice as feeding, resting and hiding places, implying an ability to survive across a range of salinities.

e On a misty Kamchatka morning, a Brown Bear and several Slaty-backed Gulls search for breakfast at low tide.
f Arctic copepods. L to R are Metridia longa *c.2.5mm, the very small* Oithona similis *c.0.5mm,* Calanus glacialis *c.4mm and* C. hyperboreus *c.7mm.*
g The amphipod Gammarus wilkitzkii *which sits head down in a brine channel waiting for prey, usually copepods, to pass.*

In more southerly waters the Greenland Halibut (*Reinhardtius hippoglossoides*) and the Pacific Halibut (*Hippoglossus stenolepis*) are commercially important. Other southerly species include the Capelin (*Mallotis villosus*), herring (*Clupea* sp.) and pollock (particularly the Alaskan or Walleye Pollock). There are also several eel species (particularly ammodytids or sandeels – sandlances in North America – which are the regular prey of auks).

But perhaps the most famous Arctic fishes, particularly on the Pacific side, are the salmonids. Adult fish are pelagic, but swim to freshwater streams to mate, lay eggs and die. Young fish reverse the journey. The salmon's spawning journey is one of the most remarkable in the natural world: some King Salmon (*Oncorhychus tshawytscha*) travel c.2,000km to reach their spawning grounds in the Yukon River, the spawning runs are a famous feature of the Pacific Arctic and are an important food resource, particularly for Brown Bears. Seven salmon species spawn in the rivers of Alaska and eastern Russia: all undergo changes in their appearance prior to spawning. Sockeye (*Oncorhychus nerka*) change from silver-blue to bright red (Photo 297c), while the male Pink Salmon (*O. gorbuscha*: Photo 297b) develops a humped back, hooked jaw and enlarged teeth. The purpose of the changes is not known: nothing similar is seen in Atlantic Salmon (*Salmo salar*).

Arctic seas are also home to jellyfish, sea anemones, sea urchins, sponges, starfish etc. Perhaps most delightful are the sea angels, small, swimming sea slugs. There is even a shark, the Greenland Shark (*Somniosus microcephalus*) which grows to 7m. It is a slow fish that feeds mostly on crabs and jellyfish, and is mainly seen in east Canadian/west Greenlandic waters (but has been seen off Iceland, Svalbard and the White Sea). One curiosity is the presence on almost all female sharks of a parasitic copepod, *Ommatokoita elongata*, which attaches itself to the fish's cornea. The parasites, which can grow to 3cm long and hang from the eye like a pink worm, feeds on corneal surface cells and ultimately cause lesions that cloud the fish's eyesight. Some scientists suggest the parasite and the shark are an example of a mutualism, the copepod being highly visible and perhaps even luminescent and so attracting prey to the fish. As the shark is thought to hunt mainly by smell in the murky waters of the deep Arctic seas, the partial loss of vision may be a relatively minor handicap. Photo 297i is an illustration of the shark and parasite from William Scoresby's *An Account of the Arctic Regions* (1820).

Freshwater rivers and lakes are, of course, also home to aquatic species. The deltas formed where the huge continental rivers reach the Arctic Ocean are a habitat for wildfowl and shorebirds, feeding on invertebrates etc. which live among the deltas' aquatic vegetation. There are few Arctic aquatic insects, though many insects with terrestrial adult forms have aquatic young – caddis flies, black flies, mayflies, stoneflies, midges and mosquitoes. Freshwater fish have also helped an understanding of Earth's glacial periods. The Lake Whitefish (*Coregonis clupeaformis*: Photo 297j top) is found in both the Mackenzie and Yukon rivers, despite the fact that the two are separated by a substantial mass of high land. It seems that two proglacial lakes formed as the ice sheet retreated north-eastwards: the lakes were once connected by a channel that allowed the fish free passage. The Blackfish (*Dallia pectoralis*: Photo 297j bottom), has an even more curious distribution, being found in the rivers of Chukotka, west Alaska and on northern Bering Sea islands. The islands were once hills rising above the plateau of Beringia. The plateau became the seabed when the ice melted, but by then the fish had populated all Beringia's rivers.

Of the fish identified in Arctic freshwaters the most northerly is the Arctic Char (*Salvelinus alpinus*), a population of which lives in Ellesmere's Lake Hazen at 82.5°N. Other species include the brilliantly coloured Arctic Grayling (*Thymallus arcticus*: Photo 297k). The Grayling is circumpolar in distribution, as are Burbot (*Lota lota*) and Northern Pike (*Esox lucius*), though in all cases the isolation of rivers and lakes means that subspecies with marked differences have arisen. Other freshwater fish include trout, minnow, carp and perch species. During the winter river fish can migrate towards the sea to reach a section of river that remains unfrozen. Fish in large lakes survive below the ice, though those in smaller bodies of water or in small streams that freeze entirely cannot survive.

*d Sea Anenomes (*Urticina eques*) and a Hyas toad crab, Svalbard.*
Sea Slugs (e) a Sea Angel and (f) Coryphella fusca, *Bering Sea.*
*f Clown nudibranch (*Triopha catalinae*), Bering Sea.*
*g Moon Jellyfish (*Aurelia aurita*), north Pacific.*
h A Branchioma polychaetes, Svalbard.

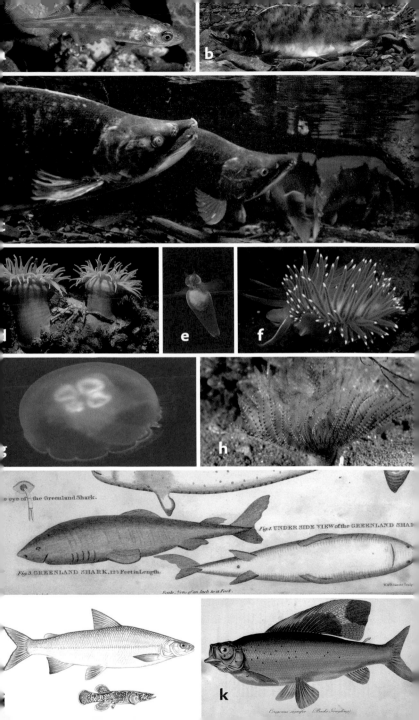

Invertebrates

Although most Arctic invertebrates are aquatic (mainly marine), a surprising number of terrestrial species inhabit both the taiga edge and the tundra. In terms of abundance the most numerous are worms (particularly nematodes and oligochaetes (Photo 299a, with dark blue springtails beneath a stone in south Greenland) and rotifers, with planarian worms being important in the taiga. Though there are marine and freshwater molluscs, there are no Arctic slugs or snails. Spiders occur as far as northern Ellesmere: indeed, several species from there are endemic, suggesting an ice age refuge that allowed speciation. Many of these spiders are very small, but on the tundra the traps of funnel web spiders can occasionally be seen, while the larger wolf spiders (Photo 299b: *Pardosa* sp. wolf spider, Victoria Island) are among the more likely specimens to be encountered. Spiders are also important on the taiga.

Insects

Insects are the most numerous of Arctic animals despite the number of species representing only about 0.3% of Earth's known species, the densities sometimes reaching staggering proportions. On dry tundra in Svalbard the density of springtails was found to be almost 40,000/m²: on damp tundra this density rose to more than 240,000/m². But for larger insects the densities decline sharply. Almost all insect orders have Arctic representatives, though there are few beetles. The most successful species are the non-biting midges or chironomids, these being as much as 25% of the total insect population in areas of the High Arctic. Many insect species are circumpolar – as many as 80% of mosquitoes and nymphalid butterflies – while many of the others show distinct trans-Atlantic or trans-Beringian ranges. Many insects are nectar and/or pollen feeders, and are important as pollinators and food for birds. They can, however, also have a negative effect. Because sections of taiga can effectively be a monoculture, any outbreak of insects that feed on the trees can swiftly reach epidemic proportions with trees being killed by the insect horde (or by microorganisms that they introduce). As well as herbivorous insects there are also species that feed on carcasses, dung feeders, and a number of predatory species. There are also significant numbers of parasitoid wasps and flies. The hosts of these insects are often home to the larvae of two or more parasitoid species.

Dipterans (flies) represent about half of all insect species, and there are representatives of the order in the High Arctic wherever their 'normal' larval habitats of dung, carrion and other detritus are available. Of particular interest to the Arctic traveller are mosquitoes and black flies, the females of which are facultative bloodsuckers, i.e. they will feed on blood if the opportunity presents itself. Male mosquitoes feed exclusively on nectar. Females feed on nectar to obtain the energy required to fly, but seek a blood meal to provide the nutrients for producing abundant, healthy eggs. Females who do not obtain a blood meal may also lay eggs, but these will be far fewer in number, and they may also produce eggs autogenously, using food reserves accumulated while they were larvae.

Mosquitoes are capable of turning joyous Arctic days, those windless hours of clear light and sunshine, into a battle to stay sane. The Inuit have an expression for such days, referring to the hordes as *sordlo pujok*, 'like smoke'. It is an expression which is not so very far from truth as the normal mosquito density of 1/m² rises to over 1,000/m², swarms of legendary size enveloping the traveller. In 1576 Sir Martin Frobisher, leader of a British expedition seeking the North-West Passage, encountered mosquitoes on Baffin Island, writing of '*a small fly or gnat that stingeth and offendeth so fiercely that the place where they bite shortly after swelleth and itcheth very sore*'. As a description it can hardly be bettered. Female mosquitoes that find a host may consume up to five times their own body weight in a single blood meal. The insect injects saliva into the host's blood to prevent clotting: it is this that causes the swelling and irritation.

c Thanatus arcticus *spider on Mountain Avens, east Greenland.*
d *The fly* Spiligona sanctipauli *on Entire-leaved Avens, west Greenland.*
e *The dung fly* Scatophaga furcata *on Entire-leaved Avens, west Greenland.*
g *Blow flies (*Calliphoridae *sp.) on a dead Musk Ox, northern Canada.*

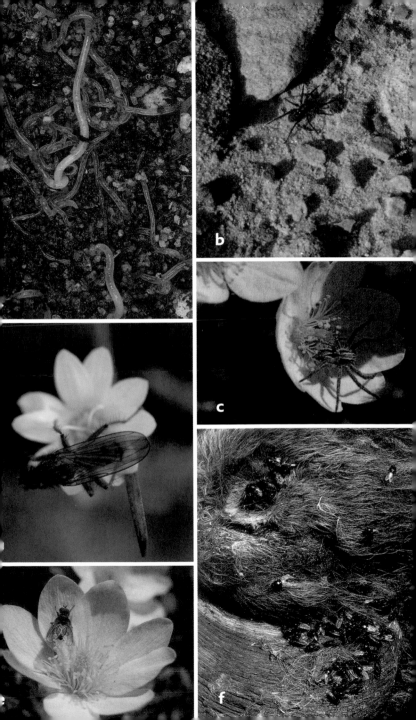

Female mosquitoes feed on any warm-blooded animal (Photos 301a, human, and b, Snowy Owl chick). The great herds of Caribou in the Nearctic are driven, it sometimes seems, to the point of madness by their attention. Although it is rare, cases are known of Caribou dying from blood loss due to mosquito bites. Birds have also been known to succumb: researchers at a Brünnich's Guillemot colony during a particularly warm spring noted the deaths of many birds from blood loss, mosquitoes attacking the feet where blood vessels were close to the surface to aid heat loss. It has been calculated that a naked human making no effort to protect himself would die from blood loss within a day. The main mosquito species in the region are *Aedes impiger* and *A. nigripes*, which are widespread with ranges that extend into the High Arctic. There are, however, about a dozen species of Arctic mosquito. The insects are drawn to exhaled carbon dioxide, and, when close to a victim, to body heat, but the claims of some travellers that they are more prone to attack than companions has some backing in science, studies suggesting that the insects preferentially visit 'healthy' victims (as 'unhealthy' people would have relatively fewer nutrients in their blood and therefore represent a poorer investment?) though exactly how a mosquito decides on fitness is difficult to understand. It is also known that mosquitoes preferentially bite pregnant women.

Black flies (*Simuliidae* sp.) also bite. They are numerous in the taiga/taiga edge where bites produce a more damaging wound than those of the smaller assassin. Thankfully they are less common and less damaging in the High Arctic. There the females do not suck blood: their mouthparts (and those of males) do not fully form. Instead eggs are produced solely using reserves built up during the female's larval stages. In the Nearctic taiga the famed No-see-ums, Ceratopogonid biting midges, are another dipteran scourge, able, it seems, to penetrate the tightest mosquito netting. Other flies are a scourge for Arctic animals rather than travellers, Reindeer/Caribou having to contend with two highly specialised fly parasites. The Caribou Warble Fly *Hypoderma tarandi* (Photos 301d, adult, and e, larva) lays a sticky egg on the legs or underside of the animal. Hatched larvae burrow into the animal, migrating subcutaneously to the back, close to the spine and excavating a breathing hole before feeding on the host's tissues. Most Caribou are infested: some carry up to 2,000 larvae, their skins being useless to native hunters as a result. The larvae overwinter in the animal then emerge through the breathing holes in spring, falling to the ground to pupate. To add to the deer's suffering, the breathing holes may become infected. The Caribou Nose Bot Fly *Cephenomyia trompe* deposits live larvae (that hatch inside their mother) at the entrance to the host's nostrils. The larvae migrate to the opening of the throat where they cluster. The larval mass can be so large that it interferes with the deer's breathing: the coughing heard in groups of deer is usually caused by animals attempting to dislodge the parasite mass: only when the larvae are ready to pupate can the deer manage to expel the mass. Both warble and bot flies are stronger fliers than mosquitoes and so are more difficult for the deer to evade. The animals lower and shake their heads when a fly is seen, and the apparently random jump and run of an individual animal is usually a sign that one of the flies has been spotted.

But not all Arctic insects are so destructive. The High Arctic has no mayflies, stoneflies or dragonflies (Photo 301f, Taiga Blue Damselfly (*Coenagrion resolutum*), Potter Marsh, Alaska), though these occur in the Low Arctic, as do grasshoppers (few in number and found only at the Arctic fringe). There are about 15 species of caddis fly in the Low Arctic, but of these only *Apatana zonella* can be termed a High Arctic species as well. Of the hymenopterans, there are stingless ants on the tundra, and some Arctic bees, although there are only two species in the High Arctic. Each is large and uses shivering to raise body temperature, the muscle mass required for this process, together with dense insulating hair, explaining their large size, which usually comes as a surprise to the first-time Arctic traveller. The queen of *Bombus polaris* (Photo 301c) overwinters, having been fertilised during the summer. In spring she founds a new colony, laying two batches of eggs. The first may contain workers, but sometimes does not, but both the first and second batches contain fertile young – new queens and male drones. In some colonies the absence of workers means that the queen herself must forage in order to feed the colony in its early stages. The second, larger, bumblebee species, *B. hyperboreus*, is a social parasite. In this species the queen also overwinters, but emerging in the spring she seeks out a *B. polaris* nest site. Bypassing workers (probably using chemical secretions), she finds and kills the *B. polaris* queen and lays her own eggs. Coated with chemicals that trick the *polaris* workers into treating them as their own. *Hyperboreus* larvae are queens, and a new year of cuckoo-like social parasitism with a twist begins.

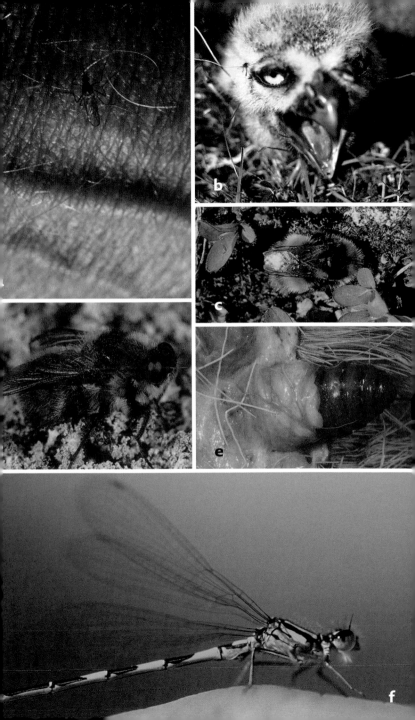

Butterflies

Butterflies are one of the surprises and joys of the Arctic summer. They can be seen as far north as flowers bloom (about 82°N). Despite their apparent fragility, butterflies are adept at using their wings as sun-collectors to raise their body temperature. Although there are more moths than butterflies in the Arctic they represent a smaller fraction of the resident lepidopterans than is usual in more temperate areas. Of the 90 or so moth species so far identified, most are micro-moths of the families Tortricidae, Noctuidae, Geometridae and Lymantriidae. The caterpillar of the Arctic Woolly Bear Moth *Gynaephora groenlandica* (of northern Greenland and Canada) takes 7-14 years to develop before pupating so long and extreme (-60°C) are the winters it has to endure (Photo 303a).

In what follows, I have tried to unpick the various common names across the Arctic countries, but the text comes with the caution that even scientific names cannot always be relied upon as subtle differences mean species are assumed when it seems subspecies are more likely. Of High Arctic species, the taxonomy of the *Boloria/Clossiana* nymphalids is most problematic, as some species and areas are poorly studied and, consequently there are species recognised by Russian lepidopterists that are not generally recognised elsewhere. Further south the American blues and the 'Spring Azure' complex, and how these relate to Eurasian blues and *Celestrina*s is also difficult. It is believed that there are no endemic butterflies of Iceland and east Greenland, those seen having been accidentally introduced by settlers. Greenlandic butterflies are concentrated in the north-west, close to Canada.

Of the butterfly families, Nymphalidae and Pieridae are the true Arctic dwellers, with blues (Lycaenidae) at lower latitudes, and skippers (Hesperiidae) and Swallowtails (Papilionidae) prominent at the Arctic fringe. High Arctic species are invariably darker than southern cousins, an adaptation to allow greater heat absorption from solar radiation. All Arctic butterflies overwinter as larvae or pupae, with some requiring several years to develop sufficiently to pupate. The most northerly of the butterflies are circumpolar. Of the nymphalids the Polar (or Polaris) Fritillary *Boloria (or Clossiana) polaris* (Photo 303b) is found on the tundra of Arctic islands to Ellesmere, on north-west Greenland and across northern Eurasia. The larvae feed primarily on Mountain Avens. The Arctic Fritillary *B. (or C.) chariclea* (Photo 303c) has a similar distribution, though it is not found as far north on Canada's islands and is uncommon in Fennoscandia. The larvae feed on Arctic Willow. Of other nymphalids, the Dusky-winged Fritillary *B. improba* breeds on Novaya Zemlya and Canada's southern Arctic islands. Frejya's Fritillary *B. freija* has a similar distribution and bears a marked similarity to the Pearl-bordered Fritillary *C. euphrosyne* (Photo 303d) of northern Scandinavia, northern Russia and Kamchatka.

Of the pierids the Northern Clouded Yellow (Hecla Sulphur) *Colias hecla* (Photo 303e) is found to northern Ellesmere and in north-western Greenland as well as across Arctic Eurasia. The Pale Arctic Clouded Yellow (Labrador Sulphur) *C. nastes* occurs more southerly on Canada's Arctic islands, but has a similar range in Eurasia. The American Clouded Sulphur *C. philodice* (Photo 303f) is found in northern Alaska.

Of the Lycaenidae, the Small Copper (American Copper) *Lycaena phlaeas* (Photo 303h), Idas (Northern) Blue *Plebejus idas* (Photo 303g), and Glandon (Arctic) Blue *Agriades glandon* (Photo 303i) are circumpolar, but more southerly while, as noted above, the Spring Azure *Celestrina ladon* of North America are related in some way with the Celestrinas of Scandinavia and Russia (e.g. the Holly Blue *Celestrina argiolus*) and others of Siberia. Of the Hesperiidae skippers, the Northern Grizzled Skipper *Pyrgus centaureae*, Chequered (Arctic) Skipper *Carterocephalus palaemon*, and Silver-spotted (Common Branded) Skipper *Hesperia comma* are circumpolar.

Of the Satyridea there are debates about the relationship of the Erebias because of the very obvious similarities (both across the species and across the continents) between the Alpines of the Nearctic and the Palearctic ringlets.

Of larger butterflies the Camberwell Beauty (Mourning Cloak) *Nymphalis antiopa* (Photo 303j) is rare everywhere, but may be seen at the treeline throughout the Arctic, while the exquisite Canadian Tiger Swallowtail *Papilio canadensis* (Photo 303k), is the only truly Arctic papilionid, breeding near the Mackenzie delta as well as in southern Alaska.

Fungi and Plants

The total number of vascular plants identified within the Arctic is around 1,000. When subspecies are taken into consideration that number rises closer to 2,000. There are also many hundreds of embryophites (mosses and liverworts), more than 1,000 lichens and a surprising number of fungi. Most Arctic vascular plants are tundra-specific, though some are also found in the taiga and to the south. About 60% of tundra vascular plants are circumpolar (with subspecies), this number rising to about 90% for the polar desert. Non-vascular plants show similar percentages, but with a greater number of species. Given the number of plants and the complexity of the relationship between the various species and subspecies, no detailed description is possible here. Instead, only a general introduction to the species likely to be seen in terrestrial Arctic habitats is given, together with details on some of the more common species.

Lichens, Mosses, and Fungi

The retreat of the ice following the last Ice Age allowed windblown seeds to germinate in the fertile glacial till left behind. But even where nothing but bare rock remained lichens could make a living. Lichens are particularly important in the Arctic, providing a valuable winter food source for mammals (the Rock Tripes, *Umbilicara* sp., also helped save the first of John Franklin's overland expeditions from starvation), as well as adding splashes of colour in otherwise uniform landscapes before and after the season for flowering plants. Lichens are dual organisms, a combination of fungus and alga, the algae lying a little way below the surface of the fungal thallus (an example of a facultative mutualism, the fungus providing water and minerals to the algae in exchange for the products of photosynthesis: free forms of lichen fungi exist, but they grow more slowly). Lichens are of three forms – crustose (crusty), foliose (leaf-like) and fruticose (shrub-like), and they will grow on virtually any substrate – soil, rocks or tree bark. In the High Arctic grey and orange crustose forms that colonise rocks are frequently seen, sometimes in places where no other vegetation is visible. One of these, Map Lichen *Rhizocarpon geographicum* (named because of its Atlas-like irregularly shaped colour patches: Photo 305a) grows slowly but at a defined rate, and can be used to measure the time since the retreat of ice from an area: it is estimated that some specimens of the lichen are at least 9,000 years old. Of the fruticose forms the most famous is the inaccurately named Reindeer Moss (*Cladonia* sp., but particularly *C. milis* (Photo 305b), *C. stellaris* (Photo 305c) and *C. rangiferina (Photo 207d)*), which occasionally form extensive and dense patches in open areas at the edge of the timberline. These lichens (often called Reindeer Moss) represent as much as 90% of Reindeer winter diet and a good fraction of the summer diet as well.

Fungi other than those associated with lichens are remarkably abundant in the Arctic, the species list probably outnumbering that of vascular plants. Mycorrhizal fungi were important in the development of the Arctic as they invade the roots of plants, obtaining sugars from the host in exchange for assisting it with mineral extraction through its array of hyphae (fungal filaments). But the more obvious above-ground fungi can also be seen, even in the polar desert of the High Arctic (Photo 305g: a puffball (*Basidiomycota* sp.), Ellesmere; Photos 305h (*Russula citrinichlora*) and 305i (*Russula nana*), west Greenland). With the invertebrates that are the primary decomposers of the temperate world relatively scarce, fungi are the major agents of decomposition in the Arctic (Photo 305d: mosses and lichens are quick to exploit the leached minerals from skeletal remains, in this case red moss and lichens on a skull on Igloolik Island in Canada's Foxe Basin). Many of the fungal families familiar in more southerly latitudes have representatives north of the treeline.

Mosses and liverworts are also seen throughout the Arctic, but are especially interesting in places where they occur in hostile environments, a tribute to the tenacity of life. On the volcanic island of Jan Mayen, vascular plants grow where windblown soils have accumulated, but the island's black lava there are delightful splashes of green where the mosses and liverworts have taken hold Photos 305e and f). Perhaps even more impressive are the vast areas of *Racomitrium* mosses have enveloped the lava fields created by the Laki eruption of 1783-84. Despite it being only a little over two centuries from the eruption the mosses are now thick enough to overtop a walker's boots (Photo 305j).

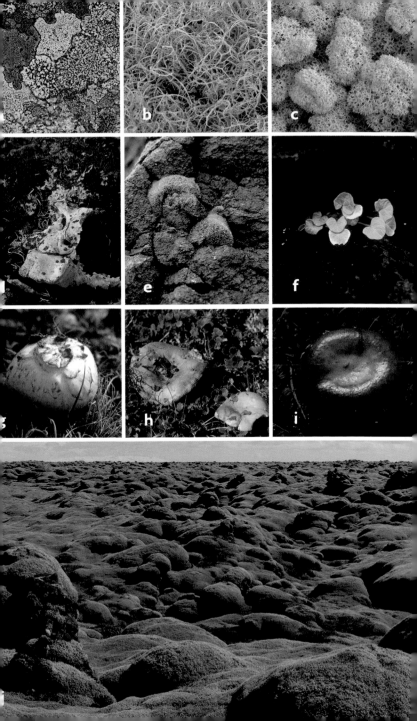

Tundra

The Finnish word *tunturia*, describing a treeless plain, has evolved into tundra, a word used to describe the circumpolar treeless belt that lies between the Arctic Ocean and the treeline. Tundra is characterised by low temperature (in general below freezing for at least half the year), low precipitation (particularly to the north) and a short growing season. But despite these characteristics applying in large part to the entire land belt between the treeline and the ocean, the tundra is not homogeneous: it has sub-divisions based on latitude and the characteristics of the landscape, each sub-division having its own vegetation.

Beneath the tundra lies permafrost, a vast and deep area over which the ground is permanently frozen, the annual thaw of the surface (the 'active' layer) providing water to compensate for low precipitation. The active layer allows plants to become established, but despite providing a growing medium, the layer is a harsh environment: it may have a negative thermocline, i.e. the temperature decreases through the layer, and it may also be waterlogged, as the underlying permafrost inhibits drainage. The depth of the active layer also defines the depth of the root structure of plants, as permafrost is as impenetrable to roots as it is to water. The layer's summer thaw is also slowed by growth of the plants it nurtures, leaf coverage preventing soil heating: in areas where vegetation is absent, the active layer may be two or three times deeper than that beneath local vegetation cover.

Attempts have been made to divide the tundra into latitude-based zones of vegetation, and while this has been reasonably successful – and are followed here – the zone boundaries are not rigid, a flexibility driven by local conditions – shelter from the wind, differences in snow accumulation, presence or absence of streams, bird cliffs, which provide fertiliser from dropping (Photo 307b: Alkhornet, Svalbard) – can mean that oases of plant life occur in otherwise unsuitable locations.

Polar Desert

The most northerly region of the tundra, covering the Arctic islands of Eurasia and most of those of Canada (though the southern parts of the most southerly islands have more extensive plant coverage), the shield area of the Canadian mainland (the Barren Lands), northern Greenland (Photo 307a) and some parts of Alaska. The notable exception to this definition is Wrangel Island which has a remarkable collection of plant species (see Mesic Tundra below).

The polar desert is cold, arid and windswept. These condition are sufficent to reduce vegetation cover, but to them must be added the long polar night and low sun angle of the polar summer which both shorten the growing season and reduce the sunlight available for photosynthesis. Low temperatures and lack of water inhibit plant growth, the wind scours exposed places and piles snow into depressions where long melt times mean that the advantages of insulation are outweighed by a further shortening of the growing season for emerging plants. As a consequence there are only scattered patches of vascular plants, the ground cover often being less than 5%, this lack of ground cover producing a further problem, a form of 'negative feedback', fewer plants meaning an individual plant is offered less protection against the elements by its neighbours and so finds survival more difficult.

The polar desert is a place for only the hardiest of plants survive – Saxifraga, Papaver, Cerastium, Dryas and Draba. *Saxifrage* derives from the Latin 'stone-breaker', which has led to the occasional suggestion that the plants aid the production of soil in the often stony alpine terrain or tundra they prefer, but this is not correct, the name actually refering to the similarity of its reproductive buds to kidney stones, a similarity that once led to the use of the plant as a remedy for dispersing the stones. The saxifrages include Purple Saxifrage (*S. oppositifolia*), which grows to 83°N in northern Greenland, making it probably the most northerly flowering plant. Competition for this title comes from Moss Campion (*Silene acaulis*) and perhaps from Mountain Avens (*Dryas octopetala*). (Photos 307c Purple Saxifrage, Igloolik Island; 'd' Moss Campion, Iceland; 'e' Mountains Avens, Ellesmere Island; 'f' Tufted Saxifrage, Svalbard).

Note that Mountain Avens can be D. *integrifolia* or D. *octopetala* as there are two closely-related species whose ranges overlap in eastern Siberia, western North America and northern Greenland. In North America the two plants are both called Mountain Avens, with the two differentiated by being given a second name (White Dryas and Eight-petalled Dryas). In limited areas of North America and in north-west and north-east Greenland hybrids of the two species

have been identified. Just to add further confusion, the Eight-petalled Dryas occasionally has seven or nine petals.

The Papaver poppies are among the most delightful of northern flowers, the long stems making them look particularly fragile. Arctic Poppy *P. radicatum* can be sulphur-yellow, but also white or even pale pink. As with other northern species, although the plant is essentially circumpolar, closely related species (which may actually be subspecies) are recognised in certain areas – Svalbard (Photo 309a) and Arctic poppies on Igloolik Island (307b); north-east Greenland (307c and d); Ellesmere Island (307e); and Baffin island (307f).

Many of the other polar desert plants are also circumpolar, but with subspecies. Draba, the white and yellow Whitlow-grasses, and *Cerastium* (mouse-ears and the related chickweeds) are particularly difficult to differentiate for the non-expert. Other plant genera represented in the tundra of the extreme north include *Ranunculus* (e.g. Snow Buttercup *R. nivalis*, Lapland Buttercup *R. lapponicus* and Arctic Buttercup *R. hyperboreus*, each of which is circumpolar. As with the northern poppies, the long-stemmed, fragile appearance of the buttercups is in sharp contrast to their actual hardiness. *Potentilla* (cinquefoils) (Photo 309g Hairy Cinquefoil *Potentilla villosula*, Alaska North Slope) and *Minuartia* (sandworts) may also be seen, as may species such as the Northern Jacob's Ladder *Polemonium boreale* (Photo 309h).

In sheltered spots the northern dwarf willows occur – Arctic Willow *Salix arctica*, a Chukotka and North American shrub (Photo 309j, Igloolik Island), and Polar Willow *S. polaris*, which is found in the European Arctic. The difference between the two is marginal and they may well be subspecies, as may the Northern Willow *S. glauca*, of northern Greenland. In more exposed areas the willow is more of a creeping woody plant, often barely more than 2 or 3cm high. Yet it is a true tree, its leaves changing to a beautiful red in autumn (Photo 309k, north-east Greenland). Arctic Bell-Heather (Arctic White Heather) *Cassiope tetragona* may also be found in sheltered spots (Photo 309i, Ellesmere Island). It too is a circumpolar species, common throughout the Neartic polar deserts, on Svalbard and Greenland: in the latter it was important as a fuel for the local Inuit.

Miniature grasses occur both inland (*Poa* spp. and *Festuca* spp.) and close to the coast (*Puccinellia* spp), and despite the region's aridity there are also sedges, particularly the drought-tolerant Cushion Sedge *Carex nardina* and Rock Sedge *C. rupestris*. In wetter areas such as stream valleys there may be Arctic Sedge *C. stans*.

Occasionally within the polar desert there are areas of exceptional plant vitality, akin to the oases in the more familiar hot deserts. These exist where local topography allows a good, well drained soil to develop in a spot where those plants that manage to take root are protected from the wind. Such an area is the valley of Ellesmere Island's Lake Hazen, at about 82.5°N, where more than 100 flowering plant species have been identified.

Southern tundra

To the south of the polar desert, tundra vegetation covers a greater percentage of the land, though initially the list of species is much the same. Ultimately more species appear – close to the treeline there are about four times the number of vascular plants than seen in the High Arctic. Some of these are shrubby, a fact that has led to the suggestion that in addition to a treeline there is also a shrubline. The height of the shrubs, and of some other plants, also increases as the climate becomes, relatively, more benign. But within this graded approach to the treeline there are specific forms of tundra in which differing species dominate.

Dry tundra or fell fields

Both names are frequently used for this tundra type, the latter deriving from the Scandinavian *fjell*, mountain. Dry tundra is an area of poor soil, a rocky or stony habitat, often exposed and so with a limited number of vascular plants, most of which maintain a low form to avoid dessication. Dry tundra is common in the southern areas of Canada's southerly Arctic islands, Russia's Taimyr Peninsula, and other upland Arctic areas. Because dry tundra tends to be windswept and so has limited snow cover, it is an important winter feeding area for Musk Ox. The plant species are similar to those of the polar deserts – Dryas, Potentilla, Draba, Salix – but sometimes with a more extensive coverage.

There are often species of Oxytropis (the oxytropes and crazyweeds of North America, and the milk-vetches of Eurasia – Photo 311h is Field Oxytrope (*Oxytropis campestris*): though normally yellow, the species is occasionally white, tinged with purple, as with this plant, which is probably a subspecies, thriving at Cape Alexander, Nunavut). Photo 311i is Northern Sweet-Vetch (*Hedysarum boreale*), Johansen Bay, Cambridge Island).

In upland areas Arctic Arnica (*Arnic alpina*) – Photo 311a: north-east Greenland – is added to the list of polar desert species. In more southerly areas, patches of matted Alpine Azalea (*Loiseleurisa procumbens*) add a splash of colour, as do Lousewarts, particularly the vivid pinks of Woolly Lousewort (*Pedicularis lanata* – Photo 311j: St Lawrence island) and Sudeten Lousewort (*P. sudetica*) while the Black Bearberry (*Arctostaphylos alpina*) – Photo 311g: northern NWT – adds a valuable berry to the diet of Arctic wildlife. Lapland Diapensia (*Diapensia lapponica*), another species that forms low cushions, also occurs here.

Dry tundra also occurs in more southerly locations, particularly those with limited snowfall. In such locations the number of berry-producing heath species increases – Vaccinium species such as Arctic Bilberry, or Blueberry (*V. uliginosum* – Photo 311f: north-west Greenland), Cranberry (*V. oxycoccos*) and Rock Cranberry (*V. vitis-idaea*), of which there are subspecies throughout the Arctic, and Crowberry (*Empetrum nigrum*), a circumpolar species which also has numerous subspecies – Photo 311e Mountain Crowberry (*E. n. hermaphroditum*): north-west Greenland. Other heath species such as Labrador Tea (*Ledum palustre* – Photo 311k: Victoria Island) and some heathers flourish, while real gems such as gentians (Photo 311d: Slender Gentian (*Gentianella tenella*): south-west Greenland) and the Arctic version of thrift can also be enjoyed – Photo 311c: Arctic Thrift (*Armeria maritima*), north-west Greenland). While much more common, large assemblies of Harebells (*Campanula rotundifolia*, with local subspecies) are a delight – Photo 311b: south-west Greenland).

Mesic tundra

Mesic tundra is an intermediate form between dry tundra and the wetter sedge and tussock forms. Watered by streams of melting snow yet adequately drained, mesic tundra is home to many varieties of grasses and other flowering plants. Mesic tundra is found across eastern Canada and through the Eurasian Arctic to the Taimyr Peninsula, though east across the Asian Russian Arctic and in Alaska and Yukon it is largely replaced by tussock tundra.

On mesic tundra Dwarf Birch (*Betula nana*) – Photo 311n: autumn Fennoscandia – occurs among several varieties of willow – e.g. Woolly (*Salix lanata*) and Downy (*S. lapponum*) Willows. In more southerly mesic areas, alders (e.g. *Alnus crispa* and *A. fruticosa*) occur. Berry-producing plants and heaths also thrive, with additional circumpolar species such as Arctic Bramble or Dwarf Raspberry (*Rubus arcticus*) and Cloudberry (*R. chamaemorus*): the latter is particularly common in northern Fennoscandia where it is much sought after, often being seen as an addition to both main meals and desserts as well as attracting many town dwellers into the wilderness when the berries are ripe. Damper areas are home to mosses and sedges.

Some authorities recognise another form of tundra, dwarf-shrub tundra, as being intermediate between dry tundra and mesic tundra. They place this form on well-drained soils, usually close to rivers or in areas with limited snow fall. However, the vegetation species list for such areas is largely as that listed above – birch and willows, berry-producing plants and species such as Labrador Tea (Photo 311k), Arctic Rhododendron (*Rhododendron lapponicum* – including the Kamchatka Rhododendron (*R. camschaticum*): Photo 311m: southern Kamchatka – and Lapland Diapensia (with Matted Cassiope (*Cassiope hypnoides*) more common in eastern Canada, Greenland and Fennoscandia). Dwarf-shrub tundra is considered to be prevalent in western Alaska, Fennoscandia and eastern Chukotka. In those areas it is an important feeding ground for Reindeer and for Snow Sheep in Chukotka, particularly as fruticose lichens often thrive among the shrubs.

Although it is one of Russia's Arctic islands, the dominant habitat on Wrangel Island is mesic tundra, with a remarkable collection of around 400 species of plant. The island was not covered by an ice sheet during the last Ice Age, and it is thought that this variety has developed as a consequence not only of this lack of glaciation, which allowed an existing flora to flourish, but from occasional periods of attachment to Beringia that would have allowed the spread of southern species. Periods of isolation from Beringia allowed the development of endemics, of which the island has many.

Wet tundra covers around half of northern Siberia, large areas of the central Canadian mainland, much of northern Alaska and areas of Greenland. In many places in the southern tundra it is the predominant form, and though it is an excellent habitat for waders and waterfowl it is rather less welcomed by the Arctic traveller. In wet tundra the dominant plant species are the Common Cottongrass (*Eriophorum angustifolium*), a circumpolar species, together with the Harestail Cottongrass (*E. vaginatum*) and White Cottongrass (*E. scheuchzeri*), sedges such as Arctic Sedge (*Carex stans*), Mountain Bog Sedge (*C. rariflora*) and Water Sedge (*C. aquatilis*). In western Siberia the dominant sedge is a particular subspecies, *C. ensifolia arctisibirica*, and Horsetails (*Equisetum* sp. – Photo 313b) the absence of which signifies the transition to east Siberia for Russian scientists. As well as this relatively abrupt switch there are also more local changes. For example, on the southern island of Novaya Zemlya and on adjacent Vaygach Island, Shortleaf Hairgrass (*Deschampsia brevifolia*) dominates. In general, either the cottongrasses or the sedges dominate in any one particular area of wet tundra. Where cottongrass dominates the white, fluffy seed heads create one of the Arctic's most aesthetically pleasing sights (Photo 313a). Mixed with the cottongrasses and sedges are grasses such as Arctic Marsh Grass (*Arctophila fulva*) and mosses. Where the ground is continuously waterlogged there are sphagnum mosses, while on drier ridges dwarf birch, heathland vegetation, berry-producing shrubs, saxifrages (Photo 313c, Marsh Saxifrage (*Saxifraga hirculus*), north-east Greenland), irises (Photo 313f *Iris setosa*, Kamchatka) and bisorts (Photo 313d, *Polygonum bistorta*, Chukotka).

A particular form of wet tundra is tussock tundra. This has a circumpolar distribution, but is most frequent in areas where the active layer of the permafrost is about 50cm deep. It is a feature of the Russian Arctic east of the Kolyma delta, particularly in Chukotka, and of the western North American Arctic from Alaska to the Mackenzie River. Tussocks are formed when the dead leaves of cottongrass and sedges take time to decompose in the cool and acidic waterlogged ground at the base of the plants. Dead material therefore builds up at the plant base. Eventually this material is converted to soil, and as more is added it breaks the water surface. It is then exploited by other plants, dead leaves from these being added to the base so the tussock height increases. As a micro-habitat tussocks are superb for microtines, wading birds and insects. Cottongrass tussocks are the almost exclusive habitat of the Siberian Lemming: to such an extent is the combination of a snow blanket acting as a thermal shield, protecting the rodents from very low ambient temperatures, and the relatively fresh vegetation at the base of the tusk, that the lemmings actually breed during the winter. But for the Arctic traveller tussocks are a nightmare. They are unstable, making tussock-hopping a risky means of travel, and the ground between them may be waterlogged, the water often overtopping a walker's boot. The combination makes for slow, hazardous travel, the misery compounded by the fact that tussock tundra is the ideal breeding place for mosquitoes. Tussocks can burn, the dead leaves at the tussock base making good tinder, but as the new buds of the cottongrasses and other plants are often buried deep inside the tussock they survive, so that the flames become a useful regenerator.

Forest or shrub tundra

Close to the treeline the shrubs, particularly the birch (Photo 313e, winter, Fennoscandia), willow and alder species, grow taller and further species – Populus sp. i.e. aspens and poplars – become established, creating an area of forest tundra. Interestingly, forest tundra often has fewer species than either the tundra to the north or the boreal forest to the south. In forest tundra berry-producing shrubs tend to grow taller and set more fruit, making the area particularly attractive to Reindeer. Rushes are found in the wetter areas. Forest tundra is a particular feature of eastern Russia, where there are large expanses between the tundra of the Taimyr Peninsula and the taiga, and significant expanses to the east of Taimyr, extending as far as the border with Chukotka. Some Arctic flowers are impressive. The Chocolate Lily (*Fritillaria camschatcensis* – Photo 313h) of Kamchatka and the Aleutians), a magnificent chocolate brown flower, is perhaps the finest example. There are also orchids: the circumpolar Calypso (*Calypso bulbosa* – Photo 315e) has a flower reminiscent of a masked carnival figure. Spotted (or Pink) Lady's Slipper (*Cypripedium guttatum*) has a distribution that includes the Mackenzie delta and the Aleutians, and Eurasia (though in the latter it is usually confined to areas south of 60°N). Across Arctic Eurasia there are various marsh orchids, while Kamchatka (Photo 315d, Purple

Orchid variant) and the Aleutians are home to several local species. One of these, the Bering Bog Orchid (*Platanthera tipuloides* – Photo 315b), which grows on Attu and more rarely on islands east to Unalaska, is considered to be the rarest North American orchid. It is tall (c.20cm) and has up to 20 tiny, golden-yellow flowers on a single stalk. In the same family, *Platanthera hyperborea* confounds visitors to Greenland where it, and several related species find a home in wet tundra in the south of the country (Photo 315c). As attractive as the orchids is the Chukotka Primrose (Photo 315a).

The Pasque flowers are a collection of related species of circumpolar distribution. Species include *Pulsatilla ludoviciana* (Photo 313g) of north-western North America and *P. pratensis* of Fennoscandia. Wrangel Island has its own endemic species, *Pulsatilla nuttaliona*. Visitors to Alaska and the Yukon will also see the Nootka Lupine (*Lupinus nootkatensis* – Photo 315g), which thrives on the Aleutians, the Pribilofs and in southern Alaska, and the Arctic Lupine *L. arcticus*, which occurs in northern areas of Alaska and nearby Canada. The Nootka has also been introduced to Iceland where it flourishes. Finally there are the gentians – the Snow Gentian *Gentiana nivalis* (Photo 315f) of Eurasia and the Northern Gentian *G. amarella acuta* of North America and Aleutians.

Boreal forest or taiga

Boreal forest and taiga are names given to the northern reaches of the forest belt that crosses North America and Eurasia, a broad belt characterised by a climate of long, dark, cold winters and short, cool summers. In general precipitation is low. There are, of course, variations; the Pacific coast of North America is warmer in winter and wetter overall, while inland Siberia has winters that are ferociously cold. The two words are interchangeable: boreal derives from *Boreas*, the North Wind of Greek mythology. Taiga has a much less definite origin. Some authorities suggest a Turkic word meaning a dense coniferous forest area rich in wildlife, while others see an alternative native Russian word meaning 'swamp-forest' or 'stick forest', a reference to the short, stunted form of trees at the northern forest edge.

Although the treeline is a useful construct it is not a clear-cut limit: there are no trees north of Alaska's Brooks Range, while across the border in Canada, near the Mackenzie delta, trees grow to the shore of the Beaufort Sea. From there the treeline heads south to the southern shore of Hudson Bay, then north again into Quebec and Labrador. In Eurasia, the treeline is no better behaved, being well above the Arctic Circle in Fennoscandia (because of the influence of the North Atlantic Drift), then heading south to the Circle, before turning north yet again. In the far east of Russia the treeline is almost a north-south line, with minimal encroachment on to the tundra of Chukotka. In mountain areas, where elevation adds extra complexity, the treeline is even more difficult to draw.

The North American taiga is dominated by spruce, White Spruce (*Picea glauca*) and Black Spruce (*P. mariana*) though Alder probably preceded the first spruces. In the vast Palearctic taiga, though the form remains more or less constant, the tree species change as the traveller heads east. In mainland Scandinavia, the forest is chiefly of Norway Spruce (*P. abies*) and Scots Pine (*Pinus sylvestris*). In European Russia, Norway Spruce and Siberian Spruce (*Picea obovata*) dominate. Further east, the great forests of western Siberia are predominantly Siberian Spruce, Siberian Fir (*Abies sibirica*) and Siberian Stone Pine (*Pinus sibirica*). In central and eastern Siberia, larch species dominate. Within the taiga there are also areas of bog, known by a variety of names, while in North America the most usual term is muskeg (Photo 315i), though true muskeg refers is a wetter bog, created where meltwater saturates the ground, and is, technically, not a true bog. True bogs form by the infilling of lakes and ponds so that the bottom layer of peat is formed of pondweeds and other water plants, while muskeg forms in areas of poor drainage. These northern bogs are home to fine marshland plants, including the insectivorous Sundew (*Drosera rotundifolia* – Photo 315j: Kamchatka) and Pitcher Plant (*Sarracenia purpurea*). But as with tussock tundra to the north, they are breeding grounds for mosquitoes and other biting insects, and a misery to cross.

Within the taiga a surprisingly high number of vascular plants flourish (Photo 315h: east Siberia). Most are shade- and cold-tolerant species, members of the Asteraceae (aster), Onagraceae (willowherb), Ranunculaceae (buttercup), Rosaceae (rose) and Brassicaceae (crucifers) families, as well as berry-producing shrubs.

Index